Cambridge Studies in Historical Geography 12

THE SUGAR CANE INDUSTRY

Cambridge Studies in Historical Geography

Series editors
ALAN R. H. BAKER   J. B. HARLEY   DAVID WARD

Cambridge Studies in Historical Geography encourages exploration of the philosophies, methodologies and techniques of historical geography and publishes the results of new research within all branches of the subject. It endeavours to secure the marriage of traditional scholarship with innovative approaches to problems and to sources, aiming in this way to provide a focus for the discipline and to contribute towards its development. The series is an international forum for publication in historical geography which also promotes contact with workers in cognate disciplines.

1. Period and place: research methods in historical geography. *Edited by* A. R. H. BAKER *and* M. BILLINGE
2. The historical geography of Scotland since 1707: geographical aspects of modernisation. DAVID TURNOCK
3. Historical understanding in geography: an idealist approach. LEONARD GUELKE
4. English industrial cities of the nineteenth century: a social geography. R. J. DENNIS
5. Explorations in historical geography: interpretative essays. *Edited by* A. R. H. BAKER *and* DEREK GREGORY
6. The tithe surveys of England and Wales. R. J. P. KAIN *and* H. C. PRINCE
7. Human territoriality: its theory and history. ROBERT DAVID SACK
8. The West Indies: patterns of development, culture and environmental change since 1492. DAVID WATTS
9. The iconography of landscape: essays in the symbolic representation, design and use of past environments. *Edited by* DENIS COSGROVE *and* STEPHEN DANIELS
10. Urban historical geography: recent progress in Britain and Germany. *Edited by* DIETRICH DENECKE *and* GARETH SHAW
11. An historical geography of modern Australia: the restive fringe. J. M. POWELL
12. The sugar cane industry: an historical geography from its origins to 1914. J. H. GALLOWAY
13. Poverty, ethnicity and the American city, 1840–1925: changing conceptions of the slum and the ghetto. DAVID WARD
14. Peasants, politicians and producers: the organisation of agriculture in France since 1918. M. C. CLEARY
15. The underdraining of farmland in England during the nineteenth century. A. D. M. PHILLIPS

THE SUGAR CANE INDUSTRY
An historical geography from its origins to 1914

J. H. GALLOWAY
Professor in the Department of Geography and Fellow of Victoria College, University of Toronto

The right of the
University of Cambridge
to print and sell
all manner of books
was granted by
Henry VIII in 1534.
The University has printed
and published continuously
since 1584.

CAMBRIDGE UNIVERSITY PRESS

CAMBRIDGE
NEW YORK   NEW ROCHELLE   MELBOURNE   SYDNEY

CAMBRIDGE UNIVERSITY PRESS
Cambridge, New York, Melbourne, Madrid, Cape Town, Singapore, São Paulo

Cambridge University Press
The Edinburgh Building, Cambridge CB2 2RU, UK

Published in the United States of America by Cambridge University Press, New York

www.cambridge.org
Information on this title: www.cambridge.org/9780521248532

© Cambridge University Press 1989

This publication is in copyright. Subject to statutory exception
and to the provisions of relevant collective licensing agreements,
no reproduction of any part may take place without
the written permission of Cambridge University Press.

First published 1989
This digitally printed first paperback version 2005

A catalogue record for this publication is available from the British Library

Library of Congress Cataloguing in Publication data

Galloway, J. H.
The sugar cane industry: an historical geography from its origins
to 1914/J. H. Galloway.
    p.   cm. – (Cambridge studies in historical geography: 12)
Bibliography.
Includes index.
ISBN 0 521 24853 1
1. Sugarcane industry – History.   I. Title. II. Series.
HD9100.5.G29 1989
338.4'763361'09 – dc 19   88-22823   CIP

ISBN-13 978-0-521-24853-2 hardback
ISBN-10 0-521-24853-1 hardback

ISBN-13 978-0-521-02219-4 paperback
ISBN-10 0-521-02219-3 paperback

*For my mother
and in memory of my father*

Contents

List of illustrations	*page* ix
Acknowledgments	xi
A note on statistics	xiii
Introduction: the taste for sweetness	1
1 Sugar cane and the manufacture of sugar	11
2 The Eastern origins	19

Part 1 The sugar industry in the West

3 The Mediterranean sugar industry: *c.* 700–1600	31
4 The Atlantic sugar industry: *c.* 1450–1680	48
5 The American sugar industry in the eighteenth century	84
6 The innovations of a long nineteenth century: 1790–1914	120
7 The geographical responses to the forces of change: 1790–1914	143

Part 2 The sugar industry in the East

8 Asia: *c.* 1750–1914	197
9 The Indian Ocean and Pacific colonies: 1800–1914	218
Conclusion	234
Appendix: the price of sugar	238
List of references	240
Index	259

Illustrations

Plates

1 Collecting manna: eighteenth-century Sicily	page 3
2 Sugar cane: perhaps the earliest representation in a western text	12
3 A Sicilian sugar mill of the second half of the sixteenth century	38
4 An Hispaniolan sugar factory in the sixteenth century	67
5 The three-roller mill, 1613	74
6 The Dutch with slaves work a three-roller mill in seventeenth-century Brazil	75
7 A sugar factory in the French West Indies, later seventeenth century	76
8 The ideal layout of a sugar plantation in the French West Indies	92
9 Two-roller mill, South India, early nineteenth century	199
10 Mortar and pestle, South India, early nineteenth century	200
11 Mortar and pestle, United Provinces, India, early twentieth century	203
12 Two-roller mill, China	205

Figures

2.1 The ancient Oriental sugar industry, c. 500 B.C. to A.D. 1100	26
3.1 Major sugar-producing regions of the Mediterranean, 700–1700	32
3.2 Sugar in the Mediterranean lands, 700–1600	35
4.1 The Atlantic sugar colonies, c. 1450–1680	49
4.2 Sugar production in the major colonies of the Atlantic period, 1450–1680	51
5.1 The colonial sugar industry, 1500–1800	85
5.2 Caribbean sugar exports, 1680–1800	86

5.3	Diffusion of innovations in America	97
5.4	The Jamaica train	98
5.5	Barbados as a center of innovation	101
6.1	Cane and beet sugar production, 1850–1915	132
7.1	Sugar production in the major English and French Caribbean colonies, 1800–1914	151
7.2	Sugar production in the Americas, 1800–1914	159
9.1	Sugar production in the Indian Ocean and the Pacific colonies, 1800–1914	220
A.1	Retail price of sugar, England, 1259–1950	239
A.2	Raw sugar (cost, insurance, freight) in London, 1730–1914	239

Tables

I.1	Consumption of centrifugal sugar	10
4.1	Sugar exports of Madeira as anticipated in the *alvará* of 1498	52
5.1	Population of the West Indian sugar colonies during the eighteenth century	114
6.1	Indian indentured laborers in the Caribbean, 1838–1917	127
6.2	The efficiency of sugar factories	140
8.1	Sugar exports of Java	212
8.2	Sugar exports of Malaya	215
9.1	Indian indentured laborers	222
9.2	Immigrants arriving in Hawaii, 1852–99	223

Acknowledgments

Barbados was my introduction to the tropics. One snowy March morning many years ago, Professor Theo Hills of the Department of Geography, McGill University, stopped me in the upper corridor of the Old Arts Building to ask me if I would like to spend the coming summer at McGill's Bellairs Research Institute in Barbados. How could I have refused? Some weeks later, I traveled through fields of sugar cane for the first time, from the airport to Bellairs on the St. James coast. My formal responsibilities as monitor of some climatological instruments took me literally a few minutes each day. Not having either the constitution or inclination to spend many hours on coral beaches, I found my way to the libraries and archives of the island, and so into the world of the sugar cane industry. I owe Theo Hills a great deal for setting me on such an interesting course and I am happy now to be able to acknowledge this debt. I would like to acknowledge also the kindness of two later academic advisors who encouraged me to keep to the course: Professor James J. Parsons of the University of California at Berkeley and Professor Sir Clifford Darby, formerly of University College London and now of King's College, Cambridge. The origins of this book indeed lie in my research on the historical geography of the Caribbean and Brazil. Alan Baker of Emmanuel College, Cambridge, convinced me that I should expand my interests in sugar cane from the Americas to the rest of the world, and he has been a patient, understanding editor. I would like to thank also the Master and Fellows of Emmanuel College, Cambridge, who did me the honor of electing me to a Visiting Fellowship for a sabbatical year which meant that I was able to explore from a very comfortable base the holdings of the Cambridge University Library. David Sheinin was a very helpful research assistant during the summer of 1983. The maps and diagrams were drawn by Jane Davie in the Cartography Office of the Department of Geography, University of Toronto, under the supervision of Geoff Matthews, a good friend who has been my cartographic counsellor for many years. I have received pointers, advice, hospitality from many people not immediately connected with my work. Pedro Geiger in Rio, and Manfredo and Stella Winge, formerly of Recife, now of Brasilia, have helped make visits to Brazil such a pleasure. Professor Sidney Mintz introduced me to the literature on sorghum syrup; Frank Plasil of Oak Ridge, Tennessee, a friend from high school days, seized the opportunity of a visit of mine, to take me to see Appalachian farmers actually making the syrup. At the University of Toronto, my friends W. J. Callahan, David C. Higgs, Aidan McQuillan and Ricardo Sternberg form a support club that I warmly appreciate. A word of thanks must also go to the Robarts Library of the University of Toronto and its librarians and to the Department of Geography, the University of Toronto, which has provided a stimulating academic home.

The Social Sciences and Humanities Research Council of Canada has helped me through research grants. Chapter 3 is a revised version of my 1977 paper on "The Mediterranean sugar industry" that originally appeared in the *Geographical Review*. I am grateful both to the editor of the *Geographical Review* and to the American Geographical Society, the holder of the copyright, for permission to republish. Chapter 5 incorporates in revised form my 1985 paper on "Tradition and innovation in the American sugar industry c. 1500–1800: an explanation", published in the *Annals of the Association of American Geographers*. I would like to thank the Association of American Geographers for permission to republish. The Royal Statistical Society has kindly given me permission to republish Figure A.1.

I gratefully acknowledge the permission of the following to reproduce their illustrations in this book: Plates 1, 2, 3, 7, 8, 9 and 10: the British Library; Plate 4: the Rare Books and Manuscripts Division, The New York Public Library, Astor, Lenox and Tilden Foundations; Plate 5: the Ajuda Palace Library; Plate 6: the Syndics of Cambridge University Library.

Victoria College,
University of Toronto

A note on statistics

The book contains a number of tables and graphs of sugar production. For ease of comparison, I have converted to metric tons all units of weight – Spanish *arrobas* and Portuguese *arrôbas* (very different measures), French *livres*, English tons, Dutch East Indian *pikols* and Singapore *piculs* (slightly different measures) – using the equivalents given by McCusker (1973: 621) and Furnivall (1939: xxiii). One metric ton equals 86.9 Spanish *arrobas*, 68.1 Portuguese *arrôbas*, 2,043 French *livres*, 0.984 English tons, 16.19 Dutch East Indian *pikols* and 16.53 Singapore *piculs*.

Many authors still turn to Noel Deerr (1949–50) for figures on the annual exports of the various sugar colonies. Unfortunately, Deerr frequently fails to mention his sources of information, although, given that he was writing in England, he no doubt found his data on the sugar trade of the English colonies in the Public Record Office. I have used statistics compiled by more recent scholars and have relied on Deerr when he is the only source.

Introduction: the taste for sweetness

Human beings enjoy sweet food and drink. Whether or not our appetite for sweet things is innate or acquired, it seems to have been with us in all parts of the world, at all times. Sugar brings out flavors, intensifies colors, and these qualities in themselves may explain why we like it so much, but it also has other virtues. At some time in the past, people learnt that sugar was also both a preservative and a fermenting agent, while in our own day we know it is a source of calories. It is an extraordinarily useful commodity. Throughout most of human history, sugar remained a luxury. Only after 1700, when Europeans had founded colonies in tropical America to produce sugar, did it appear on the world market in large enough quantities and at low enough prices for it to become a commonplace article of everyday use. In England, the annual per capita consumption increased twenty times during the years 1663 to 1775 (Sheridan 1974: 21) and a further five times between 1835 and 1935 (Burnett 1966: 274). The rise in prosperity over these three centuries had permitted more people to buy more sugar while the popularity of tea and coffee gave them new opportunities to use it, but the full explanation for the vast increase in the consumption of sugar in England must include the growing importance of the industrial processing of food and, even if the exact cause and effect relationships are not always clear, the changes in society brought about by the industrial revolution. We, in the western world, who are the largest per capita consumers of sugar, associate sugar with sugar cane (*Saccharum officinarum* L.), and rightly so, because for several hundred years we made our sugar from this plant, but cane sugar is not now and has never been the only source of sweetness.

Sugar – sucrose – occurs naturally in plants, and fruit juices and fruit syrups have been useful sweeteners and preservatives. Fig and date syrup were important in the Middle East and unfermented grape juice, or must, was available wherever grapes were grown. The Romans boiled must to concentrate the sweetness. When the must was reduced by half, it was known as *defrutum* and as *sapa* when reduced by two-thirds. *Passum*, or raisin wine, was even sweeter than the must concentrates. Maltose, a malt sugar made from germinating grains, was another ancient sweetener. The sap of trees may once have been a source of sugar in many parts of the world but, with two exceptions, it appears to have been abandoned early. The Chinese used the sap of the sagwire or sugar palm (*Arenga saccharifera*) as a sweetener well into historic times (Brothwell and Brothwell 1969: 73–84; Schafer 1977: 92–3, 109). In North America, the maple yields a particularly rich and abundant sap that continues to support a rural industry. The "sugaring-off" party in the first warm days of spring has achieved the status of an annual folkloric ritual in southern Quebec and northern New England; pictures of horse-drawn sleighs loaded with children and buckets of sap, and of primitive boiling houses set in the woods decorate many a greeting card and calendar in nostalgic remembrance of supposedly simpler days. The industry survives despite the high costs of production because of the distinctive, attractive maple taste of the sugar and syrup. Other sweet exudations of trees fall into the category of manna.

Manna does not have a precise definition, but according to Donkin (1980: 1) the word is chiefly, but not exclusively, applied to two composite categories of saccharine substances; (1) exudations from the branches or leaves of plants or trees, occasioned by unusually high atmospheric temperatures, or by the punctures of insects or artificial incisions, and (2) excretions of insects, either in the form of honeydew or, exceptionally, of protective cocoons. Most reports of manna have come from the desert areas of the Old World. The ecology of production is not clearly understood, and the occurrence of manna has often seemed a fortuitous or miraculous event. As types of manna differ, so do the methods of harvesting. Manna on the leaves or branches of trees can be collected by beating or shaking the trees so that the manna falls onto cloths spread on the ground. Collectors can tap the manna on the bark of trees by making small incisions to encourage the flow (Plate 1). In some parts of the world – Tibet, Southwest Asia, southern Italy – manna exudes with sufficient frequency and in sufficient quantities for it to have been marketed. Manna was valued as a medicine and appeared in the *materia medica* of societies from China to Spain. Calabrian manna was still in demand as a medicine even in the mid-years of the nineteenth century and was an important export of the region. In Southwest Asia, manna was used in the preparation of sweetmeats (Donkin 1980).

Sorghum is an Old World grass grown for grain and forage, but one variety, sweet sorghum (*Sorghum saccharatum*), is a source of syrup. The farmers of the Appalachian Mountains of the United States who still cultivate sweet sorghum in a minor way often refer to it erroneously as sugar cane. The original home of sweet sorghum is Africa, probably in the Sudan

Plate 1 Collecting manna: eighteenth-century Sicily
Source: Houel 1782–7: vol. 1, plate 32.

or Chad, from where it first spread down the east coast to Natal. It was grown around the villages for cattle feed, and the people chewed the canes to get at the sweet juice. It was taken to India, China and other parts of southern Asia, but only in the eighteenth century did Europeans become interested in its potential as a sweetener. The French were the most persistent but lost their enthusiasm over the great difficulty in converting the juice into sugar. Sweet sorghum was brought to the United States in the early 1850s from China via France and, in 1859, directly from Africa. The seeds were widely distributed by the government through the Middle West and South where it was thought the climate was most suitable. The farmers adopted the crop, and for about 100 years there was a sorghum syrup industry in the United States which at its peak produced 30 million gallons a year.

The industry took hold first in the Middle West, but since the 1870s has gradually retreated into the South. It has remained an artisanal activity, with the syrup being made on the farm with simple equipment, and much of it being consumed on the farm. It is not as versatile in its uses as cane sugar and is unable to compete in price. Its success in the South can perhaps be explained in terms of diet and poverty: the syrup accompanied well the traditional pork and sweet potato dishes of the region and it could be made at home without even the modest cash outlay that store-bought cane sugar required. In the hills of Appalachia sorghum syrup is made with astonishingly primitive equipment, and its survival now perhaps has something to do with the current interest in maintaining folk culture. A small amount enters commerce to please those who like its taste (Wigginton 1975: 424–36; Winberry 1980: 343–52).

Honey is a nutritious sweetener containing calcium, phosphates, iron, sulfur, vitamins C and B as well as sugar, and is a very good natural source of energy. It was, along with milk, a food of the gods, and most societies of mortals have recognized its value. Masai warriors lived on it during campaigns and in South America it was a staple in the diet of some Indian tribes. It can be gathered wild throughout much of the world, but beekeeping greatly improves its availability. Apiculture does have a long history, going back at least as far as fifth dynasty Egypt (2560–2420 B.C.). Honey had an important place in the cuisine of the Graeco-Roman world where it enriched cakes, sauces and dressings, and glazed hams – uses which even today sound very familiar. In the Middle Ages, honey preserved meat and fruit, and sweetened drinks. In northern Europe, it was the flavoring and fermenting agent in mead. Honey is now a sweetener of only minor importance but its flavor and the fact that it is a "natural food" ensures for it a continuing demand (Brothwell and Brothwell 1969: 74–80; Vellard 1939).

Given the lack of statistical data, it would be difficult to determine when cane sugar became the principal sweetener in any given part of the Old World but some reasonably accurate surmises can be made: in India where cane sugar was first made it may have achieved dominant status 2,000 or more years ago, but among the rural populations of western Europe its use may not have become widespread until the early eighteenth century, after, that is, Caribbean sugar had begun to arrive in appreciable quantities. In the New World, cane sugar was unknown until Columbus introduced it on his second voyage in 1493. When and wherever cane sugar became cheaply available, it easily eased its way into the local cuisine. Well-refined cane sugar is a more powerful sweetener than the ancient alternatives and being virtually pure it sweetens without imparting to food and drink any unwanted taste of honey, maple, fruit or wine. It is a better preservative than honey and, like honey, can be used to make alcoholic drinks. It is easy to store, easy to transport. Cane sugar remained unchallenged until beet sugar first came onto the market in the early nineteenth century. Sugar beet yields a refined pure sugar that is indistinguishable from refined cane sugar, and has proved to be a formidable rival. Boosted by subsidies from continental European governments, beet sugar by 1900 accounted for 65% of world sugar production but cane sugar has since recovered, and in the late 1980s beet contributed approximately 37%, cane 63%, to their combined total. Jointly, they still dominate the sweetener market although in recent years some new sweeteners have appeared on the scene that may give a good deal of competition in the future. Modern chemistry has produced the artificial substitutes for sugar such as *aspartame* that satisfy a specialized demand for calorie-free sweeteners. In some parts of Asia, saccharin is used as a cheap substitute for sugar which helps account for the low levels of sugar consumption there. High fructose corn syrup (HFCS), a maize derivative, is an interesting arrival. It is an expensive sweetener, and can be manufactured profitably only when protected by high tariffs on sugar, as in the United States. There, HFCS now substitutes for beet and cane sugar in some industrial processing of food, particularly in the manufacture of soft drinks. Crystalline fructose is now on the market, but industry analysts expect sales of only a few tens of tons by 1990 (Fry 1987: 17). Despite these inroads from the non-sucrose competition, production of both beet and cane sugar continues to increase.

The history of the use of cane sugar in the western world since the Middle Ages is well documented in trade statistics, in inventories of the possessions of the well-to-do and in cookery books. Returning Crusaders brought news of cane sugar to the nobility of northern Europe and they began to import it to use both as a medicine and as a rare costly additive to food and drink. For some time, sugar was sufficiently expensive as to be considered a suitable gift for princes to send each other, and even as late as 1513 the King of Portugal, in a rather extravagant and ostentatious mood, had his confectioners make life-size sugar effigies of the Pope and twelve cardinals that he then sent to Rome as a token of his royal esteem. Gradually, sugar moved from the medicine cabinet and guarded store-house to the kitchen and into the recipe books. A notable step in this progress occurred during the 1440s when sugar replaced honey in the making of chardequynce, a spiced quince and pear preserve and antecedent of marmalade that was served at the end of banquets (Wilson 1985: 27). Sugar was sold in loaves weighing up to 40 pounds, and citizens of comparatively modest means could buy portions of a loaf, a few pieces at a time. "I pray that you will vouch safe to send me another sugar loaf, for my old is done," wrote a prosperous fifteenth-century Norfolk lady to her husband in London (David 1977: 139). In Germany, by 1544, according to Braudel (1982: 191), the saying that "sugar spoils no dish" was already well-known. Sugar came in various qualities, reflecting the degree of refining. The whiter the sugar, the more refined or pure it was, and at its best it could be "exceeding white and sweet, glistring like snow" to quote the Elizabethan doctor, Thomas Muffett (1655), who moved in court circles where no doubt he had the opportunity to see a good deal of it. White sugar, especially when served from a silver caster, was in the sixteenth century a new way of conspicuously displaying wealth. Icing a cake was another. Sir Kenelm Digby (1669) was one of the first to describe how to do this, insisting of course on white sugar for the best results. The less well-refined, browner qualities of sugar were more widely used and Sir Hugh Plat made do with molasses in the recipe for an alcoholic drink he recommended in his *Delightes for ladies* (1609).

During the seventeenth and eighteenth centuries, the consumption of sugar in Europe continued to increase as the price fell. A very significant boost to the consumption of sugar came with the introduction of the new drinks. Lemonade was invented in Paris in 1630 (McPhee 1967: 71). Chocolate made from cocoa, an American plant, and tea and coffee from the East were all normally taken with sugar. In London, a Turkish merchant opened the first coffee house in 1652 (Drummond and Wilbraham 1939: 140), while tradition holds that the capture of huge quantities of coffee in the baggage-trains of the Turkish armies besieging Vienna in 1683 began the taste for coffee in central Europe. Despite the fashion for coffee and coffee houses, it did not become a truly popular drink in part because the making of a decent pot of coffee took more trouble than the making of a pot of tea, in part because coffee more than tea required sugar and/or milk to make it palatable (Burnett 1966: 124), but also because of expense. During the eighteenth century, the decline in the price of tea made it cheaper than either coffee or chocolate and readily accessible even to the poor. In England, tea became the drink of the masses, replacing gin which taxes and the rising cost of grain were making into a luxury, and even threatening that most traditional of drinks, beer. Home-brewing had once been widespread but by 1800 was becoming rare, and even commercial brewers had to look to their profits. In England, tea was promoted by the East India Company; elsewhere in Europe tea-drinking did not pervade society to the same extent, and in the south wine maintained its hold on habits and pocket-books. Towards the end of the eighteenth century, sugar approached the status of a staple in the English diet and was in the opinion of one authority, Mrs. Glasse, writing in *The complete confectioner* (1760), an "essential ingredient of cookery." So great in fact had the increase in consumption been that a controversy began over the effects of sugar on health. There were those who warned about the damage it did to teeth, but there were also soothers to argue that "That which preserves apples and plums, will also preserve livers and lungs" (Drummond and Wilbraham 1939: 281–2).

The decline in the price of sugar permitted people to assuage their taste for sweetness, but it also had a somber side. The industrial revolution created a new working class and gave it a new sugar-laden diet. Industrialization drew people from the countryside, from their gardens and fields, woods and streams which had provided their food, to the tenements and back-to-back houses where they had to buy what they ate. The long hours spent tending looms, mills and other machinery, by women as well as men, meant that there was less time for preparing meals at home. Food had to be cheap, easily served and it was often insufficient. One response of the English working class to this situation was to incorporate into its daily diet many cups of sweet tea, a calorie-laden stimulant that warms the body, revives the spirits and blunts the pangs of hunger but does not nourish. A second response was to abandon the careful cooking of traditional dishes in favor of cold or quickly heated, store-bought, factory-processed food. The "jam-buttie," nothing more than factory-made jam spread on a slice of factory-made bread, is a sugar-rich, high-calorie "convenience food," quickly prepared and quickly eaten. Sidney Mintz (1985: 180) has seen a dark side to this transformation of the English diet and has offered the hypothesis "that sugar and other drug foods, by provisioning, sating – and, indeed, drugging – farm and factory workers, sharply reduced the overall cost of creating and reproducing the metropolitan proletariat." To one exploited group long associated with sugar – the slaves who for so many centuries cultivated the cane – Mintz would now add another: the consumers.

Food manufacturers relied heavily on sugar both as a preservative as well as a means of bringing out flavors, and their new products that increasingly came onto the market during the nineteenth century – whether jams and marmalades, chocolate and confectionery, cake and biscuits, canned vegetables and fruits, sauces, soft drinks, relishes and ice cream – raised the sugar intake of all who consumed them. It even became the custom during the mid-years of the century to add sugar, as a yeast food, when baking bread although, according to Elizabeth David (1977: 111) this is unnecessary and "not a good practice." Industrial bakeries did indeed succeed in reducing the nutritional value of bread while at the same time causing the consumption of yet more sugar. This came through a change in the technology of milling flour in the pursuit of a uniformly white flour that would bake into white bread. The traditional method of grinding grain was between flat stones and in the milling the germ and bran was mixed with the flour. The wholemeal flour made dark "brown" or "black" bread. It was possible to separate the coarse particles of bran by "bolting" the flour through fine linen or woolen cloth and the resulting "white" flour did make an expensive "white" bread. For some years, millers had experimented with iron rollers to grind the flour, but good results only came when porcelain rollers were tried about 1870. The roller mills worked more quickly than stone mills, were easier to maintain, gave good control over milling and, perhaps most important of all, separated the germ along with the bran to give the white flour that had been so much sought after. But the germ and the bran contain mineral salts, vitamins and protein and once they are removed from flour, the resulting bread provides only starch and a small amount of protein (Drummond and Wilbraham 1939: 41, 348–53). Without sugar added in the baking and without a dollop of jam or syrup on each slice to provide calories, there was, and is, not very much point in eating this fashionable white bread. In England, after the 1870s, bread of any other kind was difficult to find. Given the changes in lifestyle brought about by the industrial revolution and given the products of the new food industry, which became affordable to more of the population as incomes rose, the huge increase in the consumption of sugar to about 50 kg. per capita in the twentieth century becomes easier to understand.

Similar forces were at work across the Atlantic. The eastern seaboard had long imported Caribbean sugar and the United States became a producer of cane sugar when it purchased Louisiana in 1804. Consumption of sugar increased in all parts of the vast country. By the 1850s sweet coffee – coffee from Brazil rather than tea from India and China – was the favorite hot drink in city parlors as in frontier homesteads, and the sweet fruit pie – that traditional component of the country cooking of honest, God-fearing settlers, the dessert that every American grandma baked – made its appearance in the Middle West, or Pie Belt, as Furnas (1969: 460–1) likes to call it. Certainly, Uncle Sam's reputation for a sweet tooth developed very strongly during the second half of the nineteenth century. The evidence is there in the cookery books (Hess and Hess 1977). Eliza Leslie, of *Miss Leslie's new cookery book* (1857), deplored the addition of sugar to recipes for breads and cakes that formerly had not called for it, but she could not stay the trend and by the end of the century sugar had found a formidable advocate in Fannie Farmer. She recommended the use of sugar in baking in the first edition of her classic, *The Boston cooking school cook book* (1896), but by the 1914 edition, the last for which she was personally responsible, she had increased the recommended amounts by 100%. She also favored the sweet salads that have become so popular in the United States and recipes for them appeared in successive editions of her book. To judge from sales – 3 million copies in twelve editions – Fannie Farmer had a keen appreciation of her compatriots' craving for sugar and understood how to help them satisfy it (Hess and Hess 1977: 60, 113–30).

The people of the western world have not only been able to satisfy their craving for sweetness but, whether through the manipulation of powerful forces in society as Mintz (1985) suggests or through their own volition, they have over-indulged. The alarms about the effects on health of this sugar-heavy diet already raised by the end of the eighteenth century have been repeated down to the present, and modern medical research has provided confirmation. Concerns about health, weight and appearance – fashion today is for the slim rather than the full-figure look – appear at last to have broken the two-century-long association between cheaper sugar and ever higher per capita consumption. Per capita consumption has begun to decline. In so far as the western world can be taken as a model, as the poorer countries of the world become more prosperous, so their populations can be expected to consume more sugar. The sweetness industry has a long future.

The key variable in accounting for differences in sugar consumption between societies is wealth, but culture, fashion and availability are also significant. The populations of richer industrial countries consume more sugar per capita than those of poorer rural countries; but within a rich country the influence of the other variables comes into play as the richer, better-educated, diet-conscious inhabitants consume less than the poorer and less well-educated. The greatest consumers of sugar, perhaps not surprisingly, are some of the producers, such as Barbados. In northern Europe, Australia and New Zealand, per capita consumption of sugar in the 1980s was above 40 kg. per annum. It was rather less than this in both North America where dietary concerns are to the fore and southern Europe, comparatively poor until recent years, and where wine has remained the popular drink. In parts of Africa and the Far East even today the annual per capita consumption of sugar is still only about 5 kg. (Table I.1).

This book is a study of the most important of the sweeteners so far used by mankind, sugar cane, which the demand for sugar has made into one of the major commercial crops of the world, one on which the prosperity of many countries depends. The scope of the book is, however, limited to one of the two basic aspects of the sugar cane industry, to production and not consumption, to the examination, that is, of the cultivation of sugar cane in the fields and the manufacture of sugar in the mills of the tropics, and not the refining and consumption of sugar in the markets of the world. The book begins with the origins of the industry in the East and traces its diffusion and evolution through the Mediterranean, where its long-lasting association with western imperialism began, to the Americas

Table I.1 *Consumption (kg. per capita) of centrifugal sugar*

Africa		Asia	
Egypt	32	Bangladesh	2
Ivory Coast	11	China	4
Libya	43	India	10
Mauritius	41	Indonesia	13
Nigeria	8	Israel	48
South Africa	41	Japan	23
Zaire	2	Saudi Arabia	37
Zimbabwe	23	Thailand	11
Americas		Europe	
Argentina	36	Denmark	42
Barbados	64	Poland	48
Brazil	50	Spain	31
Canada	40	Switzerland	29
Colombia	35	U.K.	42
Costa Rica	60	U.S.S.R.	44
Cuba	56	Oceania	
Haiti	11	Australia	51
Mexico	42	Fiji	56
United States	35	New Zealand	51

Note: some countries – India, Nigeria, Colombia – consume significant amounts of non-centrifugal sugar. See pp. 139, 234.

Source: Chen 1985: 39.

and finally back to Asia, Africa and the Pacific, where new sugar colonies were founded in a final burst of western expansionism during the nineteenth century. The year 1914 provides a convenient ending point. By that date cane sugar had met the first challenge from its serious competitor, beet sugar, and had been reestablished as the primary source of sucrose; while the Brussels Convention of 1902 marked the beginning of the regulation of the international sugar market. After the First World War, the sugar cane industry entered a new phase. This industry has had such important consequences for the land use and societies of the countries where it has been introduced, and has been responsible for the migration of millions of workers both slave and free, that this book necessarily is not only a study of the production of a vitally important commodity over many centuries but also a contribution to the historical geography of the tropical world.

1
Sugar cane and the manufacture of sugar

The genus *Saccharum* rates a disappointingly small entry in Bailey's *Manual of cultivated plants*, just a few lines among the many *Gramineae*, the huge family of grasses to which it belongs, and the terse descriptive statements give little idea of the controversy that surrounds the botanical history of sugar cane (Bailey 1961: 140). There has been disagreement over the number of species of sugar cane as well as over the time and place of domestication. In recent years, the authorities have found some area of common ground, but discussion continues and present conclusions may well be revised. There are now considered to be five species of the genus *Saccharum*: *S. barberi* Jeswiet, *S. officinarum* L., *S. robustum* Brandes and Jeswiet, ex. Grassl, *S. sinense* Roxb., and *S. spontaneum*. Both *S. robustum* and *S. spontaneum* can exist in the wild, but the other three depend on man for their propagation. The various forms of these species interbreed so the genus is a very diverse one. The origins of *Saccharum* may be in southern Asia: *S. spontaneum*, *S. barberi*, and *S. sinense* occur widely there, and the latter two were respectively the sugar canes of the early industries of India and China. *S. robustum* probably emerged in Indonesia and spread into New Guinea where *S. officinarum* evolved from it as the population selected the juicier forms with softer rinds to plant for chewing. The tougher *S. robustum* was and is used in New Guinea for fencing and roofing, but the attractions of the sweet *S. officinarum* led to its diffusion widely through the islands of the Pacific and back to Asia.

The sugar cane industry of the western world was for centuries based on a hybrid of *S. barberi* and *S. officinarum* that came through the Middle East from India and eventually was taken by Columbus to Hispaniola on his second transatlantic voyage in 1493. It had no name other than sugar cane until the late eighteenth century when European explorers of the Pacific brought other varieties of *S. officinarum* to the Caribbean. The old cane, long naturalized in the Americas, then became known as "Creole" and it was probably this cane that Linnaeus described in 1753 as *S.*

12   *The sugar cane industry*

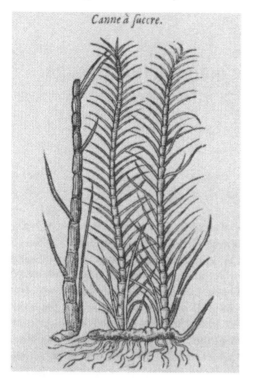

Plate 2 Sugar cane: perhaps the earliest representation in a western text
Source: Dalechamps 1615: vol. 1, 873.

officinarum L. The newly arrived Pacific varieties of *S. officinarum*, known as Bourbon or Otaheiti cane, were granted the title "noble" in testimony to their sugar-yielding properties. It was the Dutch who first talked of "noble" canes when praising the varieties of *S. officinarum* in cultivation in Java, and some of these Javan canes, particularly the Cheribon group, were widely cultivated around the world during the mid-1800s. At the end of the 1800s, researchers learnt how to breed cane, and their hybrids soon replaced those of natural occurrence in the commercial cane fields of the world (Barnes 1974: 1–5, 40–2; Chen 1985: 1–3).

The most visible part of sugar cane is its long stalk or stem which below the ground is attached to the root system and above the ground supports the leaves and the structure of the plant. The stem commonly grows to a height of 12 feet and even taller stems occur. The color of the stems may vary from yellow to green to brownish red depending on environmental conditions and varietal characteristics. The diameter of the stem too can vary in size but in mature cultivated varieties ranges normally from 3/4 of an inch to 2 inches. The cross-section of the stem is roughly cylindrical with a rind enclosing a fleshy pith. The sugar is stored in the pith cells. The rind is tough at the lower, older part of the stem but becomes softer towards the top. There is variation in the toughness of the rind and some varieties are noted for their comparatively soft rinds. The Bourbon cane, once extremely widespread, is a case in point and its soft rind explains its popularity for chewing. The stem is divided into joints which range in length from a few inches to a foot and consist of a node and section of stem or internode. The nodes are readily identifiable bands that encircle the stem, occurring more closely together at the base and top than at the center, and each contains a bud and root primordia. The significance of the nodes is in the reproduction of sugar cane.

Cultivated sugar cane is reproduced asexually, vegetatively, by planting and covering with soil sections of the stem containing at least one node. These sections are known as stem cuttings, setts or seed-pieces. New stems grow from the buds; the primary roots also grow from the nodes and feed the young stems until such time as the stems have developed their own root systems whereupon the primary roots die away. Planting fields of cane is labor-intensive work, but one planting usually gives several crops of cane. After the harvest of the first crop, known as the plant cane, the roots and lower part of the stem remain in the ground. The short section of the stem which is left below the surface of the soil has its full complement of nodes, buds and root primordia from which develop new plants while the old root system rapidly decomposes providing nourishment for the new. Each of the crops succeeding the plant cane is a "ratoon" crop – the first ratoon, second ratoon and so forth, the ultimate number depending on such factors as the local environmental conditions, the variety of cane, the yield of sugar and the incidence as well as type of pests. The yield of sugar from ratoon crops usually is less than that from the plant cane and gradually declines until the ratoons are no longer profitable. The decision when to replant will depend on the yield of sugar from the ratoon crops, the price of sugar and cost of labor. Plant cane takes from twelve to eighteen months to mature depending on the variety and the environment, although in Hawaii cane is allowed to grow for two years before harvest. The resulting heavy cane with a high sugar yield justifies the long growth period. Ratoon cane matures rather more quickly than plant cane.

The leaves of sugar cane also affect the work of the commercial cultivator. They are attached to and completely surround the stem at the nodes, and lean away from the stem above the nodes. The leaf blade is narrow and attains lengths of up to 5 to 6 feet. As the stem grows, new leaves form at the top and shade the older leaves lower down which turn yellow and fall, leaving behind them a leaf-scar on the stem. Some varieties of cane shed their leaves more easily than others. In those parts of the world where the moth borer is a pest the old leaves still clinging to the stems provide protection for the eggs and caterpillars, and water accumulating at the base of the leaves may encourage the bud and root primordia to germinate. There is no point in sending the leaves to the mill because they contain little sucrose; in fact, in the mill they will cause wastage by absorbing sucrose from the cane. The leaves should be stripped from the cane at harvest. The accumulation of dead leaves, known in the jargon of the plantations as field or cane trash, can be both an asset and a hinderance. A blanket of leaves on the ground helps conserve soil moisture for the growing crop, but dead leaves can be a fire hazard. A deliberate burn, practiced in some areas and particularly where harvesting is mechanized, clears the leaves to make way for the harvesters without loss to the sugar content of the cane, but fire in an immature crop can damage or destroy it. Practice will vary according to soil, weather conditions and labor costs.

Sugar cane will, under certain conditions, produce flowers, and some varieties are more liable to flower than others. The flowers appear on the upper terminal tip of the stem and until the late nineteenth century, they were thought to be infertile. Flowering means that the vegetative cycle is over, and the stems will not further enlarge with more sucrose. It is both unusual and unwanted in commercial canes, and where there is a tendency for it to occur steps are taken to prevent it; in experimental stations flowering is deliberately induced for the breeding of new varieties (Barnes 1974: 20–33, 267–84).

Climate strongly determines where sugar cane can be successfully cultivated. It is native to the humid tropics and demands an abundance of heat and water the year round for the best results. The threshold for satisfactory growth is 21 °C, but higher temperatures between 27 °C and 38 °C are preferable. Below 21 °C the rate of growth is greatly reduced and none at all occurs below 11 °C to 13 °C, the precise point depending on the variety of cane. In cold weather cane will not germinate. Clearly, therefore, on the outer margins of the tropics where there is a cold winter season, the growth of cane can be greatly handicapped. Even so, the risk of frost has not prevented the emergence of sugar cane industries in such subtropical regions as the southern U.S.A., southern Brazil, Argentina, Egypt and Natal. The actual effect of frost on cane depends on the variety and health of cane, and on soil moisture. In general, the more prolonged and the more intense the frost, the greater the damage. A light frost of less than an hour will cause only superficial damage to the canopy and leaves, but temperatures of −4 °C to −3 °C will kill leaves and terminal buds, and lower temperatures of even only short duration will do much more extensive damage. Some strategies can be adopted to lessen the frost risk. Where the terrain permits, as in southern Brazil, frost hollows can be avoided;

over small areas "smudging" – that is using small portable stoves to create smoke in the hope of lessening radiation – can be helpful. But there is a limit to the amount of protection that can be given and the polar range of sugar cane cultivation is set where the damage from frost and low temperatures prevent the growing of profitable crops (Barnes 1974: 328–31; Chen 1985: 15–16). Sugar cane requires an abundance of water the year round in order to keep the cane in continuous growth. In few parts of the tropics is the rainfall regime ideal, but the cane will mature in somewhat less than ideal conditions. In areas of low rainfall and/or a marked dry season, the cultivation of cane requires irrigation. The association between sugar cane and irrigation goes back several centuries, at least to the introduction of sugar cane to the Middle East; and today, thanks to knowledge of evapotranspiration, dry tropical environments are the most favorable for the commercial cultivation of cane, always provided, of course, that irrigation can ensure an optimum supply of water from planting to harvest (Blume 1985: 93). Even within the humid tropics, irrigation is now often used in order to control the growth of the crop.

Soils are a less critical environmental factor than either temperature or water. Sugar cane is tolerant of a wide range of soil types without any particular type being considered the "best." It flourishes in soils as diverse as alluvial bottom lands, the red clays of the coastal hills of Pernambuco and the coral limestone soil of Barbados. Cane does, however, make heavy demands on the fertility of soils and the application of large amounts of fertilizer has long been an important feature as well as an expense of sugar cane cultivation.

Ideal location does mean that the owner of the cane fields is free of cares, as the plant is very susceptible to disease and also is subject to attack by a wide range of animal and insect pests. Rats infest cane fields – populations of up to 250 per acre have been counted – and by gnawing on the stems damage the crop. They have been a persistent problem over the centuries but as yet no satisfactory means of control has been found. In parts of Asia and Africa, elephant and hippopotami invade the fields to make a meal of young cane. Insects bore into and feed on different parts of the cane and some spread disease. The moth borer flourishes in the cane fields of the Americas, but the extent of the damage it does varies according to the variety of cane, climate and presence of natural enemies. During the nineteenth century some outbreaks of disease were so serious as to threaten the continued existence of the industry in some locations. In Mauritius and Réunion crops suffered severely from disease in the 1860s, and in Java during the second half of the nineteenth century, Sereh disease devastated the cane. There were serious outbreaks of red rot, root rot, mosaic disease and Fiji disease in a number of countries around 1900. Damage from disease was the major impetus to begin experiments with cane breeding to evolve disease-resistant varieties. For reasons that are not yet understood, these new varieties bred by man have characteristically with the passage of time become less productive and more susceptible to disease and pests, and have had themselves to be replaced by yet newer varieties. The industry has become dependent on cane breeding (Barnes 1974: 306–43).

The sugar cane industry, in the past as well as today, divides into three distinct stages: the cultivation and harvesting of the sugar cane, the extraction of the juice from the cane and the conversion of the juice to crystalline sugar. The long history of the industry has been marked by efforts to make the processes involved in each of these three phases more efficient, and innovations in one phase have often had repercussions on the other two.

By 1914, there had been least change in the first of these stages: the planting and harvesting had always been and remained labor-intensive, hot, ardous, back-breaking toil, and as the industry grew in scope so it had demanded an increasing supply of laborers. The late nineteenth century saw the introduction of narrow-gauge railways to carry the cane from field to mill, but in the field work itself there was no respite. Not until after the First World War did mechanization begin in regions with high labor costs, and machines, difficult to design and expensive to buy, only slowly replaced men. As late as the 1960s, Fidel Castro would go out onto the cane fields of Cuba to encourage workers to get in the crop with a macho demonstration of his own prowess with a machete, and even now machinery has not replaced hand-harvesting where the terrain is difficult for machines or the wage rates are very low.

A good deal of sugar cane never goes to a mill; it is chewed, usually by the people who grow it, but this pleasant practice leads to direct consumption of the juice. For the manufacture of sugar, some sort of press or mill is required as well as a boiling house or factory where the juice can be converted into raw sugar. Because cut sugar cane deteriorates rapidly – it should be milled within twenty-four hours of the harvest – and also because it is bulky and hence expensive to transport, the mills and factories are placed among or within easy reach of the cane fields. This was as true in the past as it is today, the only difference being that improvements in transportation from carts to narrow-gauge railways to trucks have extended the distance over which mills can draw their supplies of cane. The extraction of the juice at first was carried out in a very simple manner with a mortar and pestle or by pressing two stones together, but the design of mills and presses over the centuries became more complex, reflecting human inventiveness and technological progress. From the mill, the juice flows to a reservoir in the adjacent boiling house or factory and is ready to be made into raw sugar. The manufacture of sugar from the juice can either be a simple or very complex process depending on the quality of the end product desired and the technology available. The first end product was probably a viscous residual mass of impure uncrystallized sugar, akin, presumably, to the *gur* that is still made today in northern India by boiling down the juice in a container over a fire or stove. The manufacture of crystalline sugar divides into the following steps: impurities must be removed from the juice, the juice then carefully reduced to the "strike point" when crystals form, the remaining molasses drained from the raw sugar and the sugar then dried. The quality of the raw sugar will depend on the sophistication of the manufacturing, but invariably raw sugar requires further refining, which involves dissolving and reboiling and the removal of more impurities, before it becomes pure, granular and white. It is raw sugar that enters international trade: the refineries are close to where the sugar is to be consumed.

The fundamental reason for the separation of the final stage in the manufacture of sugar – refining – from the cane fields, a separation that in the western world dates back several hundred years, lies in the fact that crystals of sugar coalesce during the humid conditions of a long sea voyage, and so any imported refined sugar would have had to have been reworked if customers were to have received the top quality. In other words, the nature of sugar dictated the location of refining, but two other considerations reinforced this solution. The making of raw sugar and the refining of sugar require large amounts of fuel. For much of the history of the industry, fuel was in short supply in many cane-growing regions to the extent that often there was difficulty in obtaining even enough fuel to make the raw sugar, and finding fuel for refining was out of the question. Finally, refining created jobs, and the metropolitan governments preferred these jobs to be at home rather than in the colonies.

Changes in technology have altered the social structure and settlement patterns of the cane-growing regions. The drive for greater efficiency has led to ever larger mills and factories, which have required in turn correspondingly larger supplies of cane to keep them operating at maximum capacity. As mills increased in size, fewer mills were needed and the greater capital costs of the larger mills restricted ownership to the wealthiest members of society. Given the necessity of coordinating the flow of cane from field to mill and factory, the millers found the ownership of cane fields gave them greater control of their supply of cane, and so bought more land. Cultivators of cane who did not have the resources to keep up with the pace of technological change had the option of continuing their activities with out-dated equipment so long as they could find a market for the sugar or of paying for their cane to be crushed by a neighbor with a modern mill. The technology-driven upward trend in the optimum size of sugar estate, mill and factory made the hierarchical plantation society more and more narrow at the apex and increased the capital required to keep up-to-date. By 1914, in many parts of the cane-growing world, large corporations were taking over ownership of the industry from the planter class.

The course of change in the industry has not been an even progress across the centuries. Rather, there have been periods of slow evolution when generation after generation of workers routinely followed traditional methods of cultivation and manufacture and others in which an individual innovation has opened up new prospects, or the innovations have come so quickly and been so inter-linked that their total effect on the industry has been revolutionary. The application of irrigation to the cultivation of sugar cane by the Arabs in the early Middle Ages, the access to the abundant resources of the Americas after 1492 and the technological and scientific advances of the nineteenth century, all in their time transformed patterns of production. Usually, for a mix of economic, cultural and environmental reasons, one region or another was foremost in adopting innovations and it is possible to identify over the centuries a sequence of centers of innovation from which knowledge disseminated around the cane-growing world. Perhaps the most important of these centers was the first, northern India.

2
The Eastern origins

In all probability, the sugar cane industry does not have a single place of origin given the simple and rather obvious nature of the processes that are the basis of the manufacture of sugar. It is reasonable to suppose that many societies in prehistoric southern Asia took the step from chewing sugar cane, whether wild or cultivated, to pressing out the juice with some kind of mechanical aid and that also they discovered for themselves the advantages of boiling down the juice into a *gur*-like viscous mass so that they could transport it more easily and use it in cooking. The case for parallel or multiple invention is not so clear for the much more complex process of making crystalline sugar, and knowledge of the techniques involved could well have diffused from some single region of discovery. So far the archaeological record provides no enlightenment as to where or when the first sugar makers began their work and indeed we may never find out as the modest scale of their activities left little to be discovered in the way of distinctive equipment or buildings by which to identify them. The first literary evidence of sugar manufacture is in Sanskrit, and so the history of the sugar cane industry begins in northern India.

Northern India

Sanskrit literature yields tantalizing glimpses rather than substantial information of the earliest years of the sugar industry, glimpses that show evidence of progress in technology but leave obscure the time of such fundamental advances in the history of the industry as the first manufacture of *gur* and the discovery of the means of purifying and crystallizing the cane juice. The evidence is linguistic, revolving around the original and new meanings of words and when the new meanings were first adopted. As the texts themselves in which these words occur often cannot be placed more accurately than to within a space of three to four centuries the fixing of dates in any chronology of events in the development of the industry

20   The sugar cane industry

must be tentative. Recent Indian agricultural historians have dealt with this difficult legacy in contrasting ways. Raychaudhuri and his associates in their *Agriculture in ancient India* (1964: 79–80) are disappointingly brief and cautious, and do not venture an opinion on dates of the innovations. Gopal (1964) by contrast is incautiously bold, making strong claims on the basis of the tendentious evidence. His paper on "Sugar-making in ancient India" is the most detailed study of the subject in a western language but his assumption of familiarity with the sources unfortunately makes it rather difficult reading for those who are not Sanskrit scholars. The discussion that follows relies on Gopal, but the dates for the various texts cited by Gopal are taken from Banerji, *A companion to Sanskrit literature* (1971).

The earliest form of Sanskrit literature, the hymns of the Vedic period (*c.* 1500–500 B.C.), contain references to the cultivation of sugar cane but not to the manufacture of sugar. For some unknown period, it can be inferred, the people of northern India grew cane for its juice which presumably they first extracted through chewing and, later, through the use of simple mills and presses. The dating of the beginning of the manufacture of sugar hinges on the etymology of two Sanskrit words *guḍa* (or *guḷa*) and *śarkarā*. *Guḍa* derives from an Indo-European root meaning "to make into a ball," "to conglomerate," and was first used in the sense of a ball; later, in Sanskrit as well as in the Pali texts of early Buddhist literature, it came to mean, according to Gopal, some sugary coagulate produced by boiling the juice, or, in other words, *gur*. The earliest Sanskrit use of *guḍa* with reference to sugar is in the work of Pāṇini, the great Sanskrit grammarian. Unfortunately for our purposes, there is no consensus among Sanskrit scholars on when Pāṇini lived other than between the seventh and fourth centuries B.C. The Pali occurrences are in the *Jātakas*, ancient folk tales that can be placed no more precisely than between the sixth to the fourth centuries B.C. There are many references to the cultivation of sugar cane and the manufacture of sugar in the *Jātakas* including the use of presses to extract the juice. The Sanskrit *śarkarā* originally meant granular particles or gravel and came also to mean crystalline sugar. Its appearance in the lexicon of the sugar industry denotes that the Indians had learnt to crystallize sugar, and as a by-product of crystallization they would also have been producing molasses. *Śarkarā* appears with this new meaning for the first time in the *Pāṇinīya* a text that probably pre-dates 500 B.C. (and may not be the work of Pāṇini). The earliest precise and secure date for the manufacture of sugar from sugar cane is in the *Arthaśāstra*, a Sanskrit manual on statesmanship written *c.* 324–300 B.C. by a government official known as Kautilya who records five distinct varieties of sugar, including notably *guḍa*, *khaṇḍa* (a word that has come down in English as candy), and *śarkarā*. Kautilya's information permits the inference that by his time manufacturers had learnt sophisticated refining techniques that permitted them to produce sugars of different degrees of purity.

Later writers, c. 500 B.C. to A.D. 1000, in the so-called classical period of Sanskrit, reveal a detailed knowledge both of the botany of sugar cane and of different qualities of sugar. Caraka, a physician of the first or second century A.D. identified two varieties of sugar cane; Amarasimha (fl. fourth century A.D.?) argued there were more varieties, while Suśruta (fl. ? but post-Caraka) described no less than twelve. These authors were noting the presence in northern India of what we know now to have been the wild *S. spontaneum*, the domesticates *S. officinarum* and *S. barberi* Jeswiet as well as quite possibly hybrids and mutants. Both Caraka and Suśruta used the same names as Kautilya to distinguish different qualities of sugar but it is difficult to know whether the varieties remained unchanged over the centuries or whether names persisted even as the qualities of sugar they described had become more refined as the process of manufacture improved. The varieties of sugar ranged in order of increasing purity from *guda* to *śarkarā*. That a doctor, Caraka, should have been so knowledgeable about sugar, itemizing as he did the medical properties of each variety, suggests that sugar was well-entrenched in the pharmacopoeia of northern India. The ability to make crystalline sugar led to an increase in demand which in turn during the succeeding centuries stimulated an extension in the area planted to sugar cane, and at the time of Gupta rule (c. A.D. 300–550) its cultivation as a commercial crop in the Ganges valley was "widespread" (Maity 1970: 109).

The Sanskrit sources thus provide the material for only a minimalist description, recording little more than the existence of a sugar industry that over the centuries experienced some improvements in the process of manufacturing. There are lacunae in our knowledge of almost every aspect of activity of this ancient industry – agricultural practices, milling technology, methods of manufacture, quantities produced, importance to the regional economy, types of labor, patterns of landownership all go unreported – and it is unlikely given the remoteness of the period and the scant chance of new sources coming to light that these lacunae will ever be filled. But the importance of northern India in the historical geography of the sugar cane industry cannot be denied: it was the first center of innovation and from here knowledge of how to make crystalline sugar spread along the trade routes to the Far East and through Iran to the West.

China

The introduction of the sugar industry to China is comparatively well-documented. In late Chou (twelfth century B.C. to 221 B.C.) and early Han (206 B.C. to A.D. 220) times the Chinese knew of sugar cane as a crop grown in Tonkin, Annam and elsewhere in Indo-China, and from there they received either through trade or in tribute such commodities as cane juice, syrup, "sugar liquor," which may have been a fermented drink, and, by the third century A.D., cakes or loaves of hard sugar formed into the shape of men or animals that were known to the Chinese as "stone honey." This Indo-Chinese industry is only known through the records of what the Chinese imported. No descriptions survive of the methods of cultivation or milling, and the single brief statement on methods of manufacture, quoted by Schafer (1963: 153), that the sugar for "stone honey" was made by "drying the juice of the cane in the sun" is surely an inaccurate description of the process by which the Tonkinese produced what was presumably a crude form of crystalline sugar set in molds. A case can perhaps be made that this industry represents a second independent discovery of techniques of sugar refining, parallel to the discovery in northern India, but as Southeast Asia lay in the path of cultural influences emanating from India it may just as well have been the result of the diffusion of ideas. Some centuries later, during T'ang times (A.D. 618–907), "stone honey" also reached China via the Silk Road across central Asia from Iran.

The cultivation of sugar cane spread into China from the south at a date that is a little uncertain, during the Han dynasty or even a little before, and by the T'ang dynasty "it was growing well in central Szechwan, northern Hupeh, and coastal Chekiang" (Schafer 1963: 152). The first references to the actual manufacture of cane sugar in China are from later Han times, and during the T'ang era sugar was produced in a number of centers. Just what the end product was and how it was made is difficult to know, but the Chinese recognized that their methods were inferior to those in use in India, for in A.D. 647 the Emperor T'ai Tsung sent a mission to study the industry in Magadha, a state in the Ganges valley in present-day Bihar. Schafer (1963: 153) suggests the mission returned with the knowledge of how to make a "rather good 'brown sugar', granular, but not truly refined" and that the ability to make a refined crystalline sugar of the sort the Chinese would call "sugar frost" was a development of the Sung dynasty (A.D. 960–1279). Nevertheless, whatever the quality of Chinese cane sugar, by the seventh century A.D. it had already gained an accepted place in Chinese cuisine alongside honey, manna, maltose and sagwire palm sugar, to be used as availability, taste and budget allowed as a seasoning in various dishes as well as in expensive confections, some of which at least involved boiling together sugar, milk and rice powder (Laufer 1919: 376–7; Li 1979: 57–9; Schafer 1963: 152–4; Schafer 1977: 108–9; Yü 1977: 57, 67).

China thus appears as the second large cultural region to adopt sugar cane as an important cash crop. There were no climatic barriers in the way of the diffusion of sugar cane from New Guinea, India and Southeast Asia into southern China through humid tropics. On the other hand, the lack of information makes it difficult to know whether social and/or economic barriers slowed the incorporation of sugar cane into Chinese agriculture and of cane sugar into the Chinese diet. Certainly, cane sugar came to the Chinese as an additional sweetener and did not immediately displace those that were already part of Chinese cuisine. Cane sugar took much longer to reach the West, but once there rapidly became the dominant sweetener and retained this position for several centuries until challenged by beet.

The way West

The journey of the sugar industry from India to the eastern Mediterranean provides a very interesting case study in cultural diffusion. The dimensions of the problem emerge from a consideration of the well-confirmed chronology of the diffusion. The few references in some Greek and Roman authors to sugar cane show that the late classical world knew of the crop and its product through trade and travelers' reports some centuries before the plant actually reached the shores of the Mediterranean. Secondly, the evidence suggests that sugar cane made the comparatively short move from India into Persia only c. A.D. 600, long after it had become an important crop in northern India, but from Persia it spread rapidly to many points in the Middle East. Finally, the commercial production of cane sugar followed the diffusion of the plant by a century and more. The discussion must therefore account for the late start and subsequent rapid pace of the diffusion, identify the agent or means of the diffusion and explain the timelag between diffusion of the plant and commercial production. In other words, the diffusion of the sugar industry from northern India to the Mediterranean can be examined as four interrelated problems each of which must be resolved if this phase in the historical geography of the industry is to be understood. The establishment of this chronology owes much to the pioneering work of von Lippmann (1929) and Deerr (1949–50) both of whom searched the classical literature and the Arab sources for evidence of the diffusion. They, and others, have recognized the fundamental importance in the diffusion of the Arab conquests after the death of Mohammed in A.D. 632 but failed to describe with precision the nature of the Arab role. It is unlikely that new archival sources on this subject will ever be forthcoming and so the one way forward is to make better use of the data that are to hand, to reinterpret them from a different perspective. Recently, Andrew Watson has exhaustively reviewed the botanical and documentary evidence of the diffusion of sugar cane to the Mediterranean as part of his study of *Agricultural innovation in the early Islamic world* (1983). Set in this larger context of agricultural change, the various stages in the diffusion westward of sugar cane are more easily understood. The paragraphs that follow rely on Watson's work.

The dates and routes of the diffusion in so far as they can accurately be determined are these. The first references to sugar in Persia occur during the Sasanian empire (A.D. 211–651). The documents suggest the rulers had a taste for sugar but they are not explicit about its origins: the sugar could have been imported and the sweet canes that are mentioned may have been sorghum. The first firm evidence of the existence of a sugar cane industry comes from the late seventh century, in Chinese and Islamic sources. Given this evidence follows immediately the Islamic conquest of the Sasanians, it is reasonable to infer the industry pre-dates the conquest by a few years, which places its beginning to *c.* A.D. 600. A tax levy shows that sugar cane was being cultivated in Mesopotamia during the years A.D. 636–44 but this is the only evidence of its cultivation in this part of the Middle East until the tenth century when numerous records suggest that it was at that time an important crop. The first undoubted reference to sugar cane in Egypt is from the mid-eighth century; more mentions follow from the ninth, and in the tenth it was grown in the Delta and Lower Nile. To generalize, the pattern of evidence indicates a diffusion of sugar cane in the seventh and eighth centuries that preceded its emergence as an important cash crop in the Middle East by the tenth. While no production figures have survived it would appear that the most important centers of the industry during the tenth century were the former Sasanian lands, the eastern shore of the Persian Gulf and in particular the region around Hormuz from which the crop may have been taken to Mesopotamia, the valleys of the Tigris and Euphrates, the lands around the southern shore of the Caspian Sea, the Damascus oasis, the valley of the Jordan, deltaic Egypt and the Lower Nile valley as well as the coastlands of the Levant.

A second route of diffusion from India led west and south to Oman, across southern Arabia and on into Africa, and as in the case of the first route it is also very poorly documented. References by Pliny and Dioscorides suggest sugar cane had reached Arabia as early as the first century A.D. but Watson is skeptical on the grounds that these authors knew little about either sugar cane or Arabia, and did not realize that the Roman traders merely collected in Arabian ports sugar that had been brought there from India. Almost certainly sugar cane was a post-Islamic introduction to Arabia, for the first firm evidence of sugar cane there dates only from the ninth century; in the tenth several writers reported it in the Hadramawt as well as in Yemen. By the tenth century, also, cane was being cultivated along the coast of East Africa, and Zanzibari sugar already had a reputation for good quality. The diffusion may have reached as far south as Madagascar. The first mention of sugar cane in Abyssinia dates from the twelfth century, although almost certainly the introduction was much earlier, but there is little indication that it spread further into the interior of the continent (Watson 1983: 24–30, 159–62) (Fig. 2.1).

The Arab role in this diffusion must now be clarified. Sugar cane was only one of several crops that spread from India through the Middle East to the Mediterranean at this time. Cotton, bananas, mango trees and colocasia – like sugar cane – were plants of humid tropical origin adapted to climates with abundant water from heavy rains, and they all required irrigation to grow in the dry climates to the west of India. The ancient Middle East did of course know about irrigation but the techniques of watering the land were adjusted to the demands of field crops that were grown during the cooler part of the year; in contrast, sugar cane and others of the new crops needed water during the summer as well. Along the Nile, for instance, the cultivation of sugar cane benefited from the annual floods, but between floods the crop had to be irrigated twenty-eight times by drawing water up out of the river. The diffusion of the new crops, in brief, depended on the development of more elaborate and sophisticated irrigation techniques. However, the barrier to the diffusion westwards from India was not only environmental; it was political, cultural and institutional as well. The new crops required landowners and laborers to learn new agricultural techniques; the construction of new irrigation schemes required a large investment of capital, a legal framework to govern the distribution of the water and a market for the produce. The market could come from rising prosperity and an increasing population which in turn depended on peaceful times and political stability. The creation of the conditions in which these agricultural improvements could be introduced was of crucial importance to the timing of the westward diffusion of crops out of India, and it is the singular contribution of the Arabs to have brought these conditions about in the years following the birth of Islam.

The missionary fervor and military conquests that within a century had brought the new religion and Arab rule to a broad swath of territory from Sind to Spain gave way to a kind of pax Islamica in which the unity of language, religion and for some periods at least even of administration made for the relatively easy movement of people and goods and knowledge. Traders and scholars, pilgrims and governors, all sorts of people it seems from all classes of life, traveled to and fro between ports and caliphal capitals observing and learning as they went, and disseminating about the Moslem world new ideas, techniques, styles of architecture and tastes in food. Another characteristic of this emerging civilization was a respect for education. The works of scholars circulated widely, rulers maintained extensive libraries and their gardens were not only places of beauty but centers of research where botanists conducted agricultural experiments and cultivated exotic plants. Scientific curiosity may, perhaps, explain the early

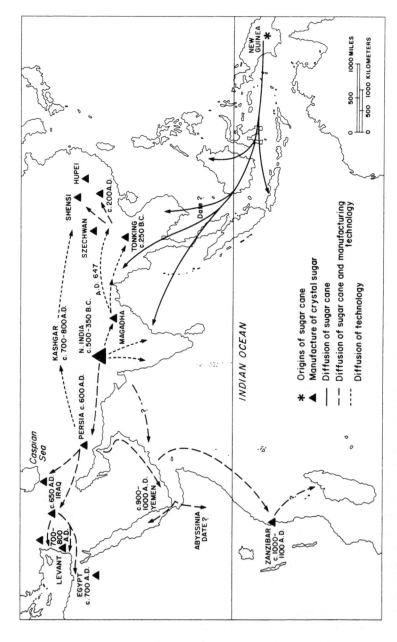

Fig. 2.1 The ancient Oriental sugar industry, c. 500 B.C. to A.D. 1100.

post-conquest diffusion westwards of the plants as evidenced in, for instance, the reports of sugar cane at various localities in the Middle East; the economic support for the conversion of these plants from rarities into cash crops of some importance came a little later as population and wealth increased, and larger and larger sections of the community could afford to indulge once-luxurious tastes. During the ninth and tenth centuries the political and economic circumstances were particularly favorable and landowners in one region after another decided to invest in the cultivation of the new crops (Watson 1983: 77–102).

The state clearly encouraged the agricultural revolution and not only through the support of botanical gardens and the circulation of agricultural manuals. One important effect of the Arab conquest was the removal of various traditional or communal restrictions on land use that inhibited change. In post-conquest times land was bought and sold on an open market and the owners could plant the crops they thought would bring the best return. Tax policies permitted owners to reap the benefits of the improvements they made and there is evidence to suggest that some of the new crops were exempt from taxation. Large-scale irrigation benefited from stable governments and good administration. According to Watson (1983: 108), "The Islamic contribution was less in the invention of new devices than in the application on a much wider scale of devices which in pre-Islamic times had been used only over limited areas and to a limited extent." The noria or water wheel, known but not widely used in the pre-Islamic Middle East, was improved and its geographic range greatly extended. Likewise, the pre-Islamic and apparently Persian (Iranian) invention of tapping the water table by means of an underground canal or *qanât* had a very limited distribution until after the Islamic conquest. The Islamic irrigation engineers became very adept at integrating various means of tapping, raising, storing and distributing water to make maximum use of the available supply (Watson 1983: 103–19, 190–200).

As a consequence of these improvements agriculture flourished in the Middle East as it never had before, and the range of the sugar industry was extended from India to the Mediterranean. In comparison to the Middle Eastern industry, the activity of cultivators and manufacturers around the shores of the Mediterranean is well-documented. In the Crusader states, Cyprus and Sicily, west Europeans first encountered sugar cane and learnt how to make sugar and forged the link between the sugar industry and European colonialism that has endured until our times.

PART 1

The sugar industry in the West

3

The Mediterranean sugar industry: c. 700–1600

For almost a thousand years North Africa, the Levant and Europe received their supplies of sugar from an industry established around the shores of the Mediterranean. This industry began about A.D. 700, flourished in many different parts of the region and finally succumbed during the sixteenth century to competition from the new plantations in the Americas (Fig. 3.1). The demands of new forms of land use and the pressures of population in these densely inhabited coastlands have over the centuries removed almost all traces of the ancient industry from the landscape. The rare reminders include a few ruins of stone sugar works in Palestine, Cyprus and in the deserts of southern Morocco; the Gate of the Sugar Workers in the walls of Syracuse, which attests to the former importance of the sugar industry in Sicily, and the few fields of sugar cane that are still cultivated, now heavily subsidized, behind the tourist beaches of southern Spain near Motril. Yet the long association of sugar cane cultivation with the Mediterranean is largely forgotten, and its place in the historical geography of the region is little known.

Historians of the medieval and early modern Mediterranean have taken very little interest in the sugar industry, seldom giving it greater mention than as an exotic item of trade and object of curiosity to Crusaders. The medievalist who is possibly the most cited on the topic is the nineteenth-century scholar Wilhelm von Heyd (1879). Even Fernand Braudel (1972–3) in the most recent edition of his magisterial work on the Mediterranean world of the sixteenth century scarcely notes the existence of the industry. This neglect can perhaps be explained by the fact that sugar cane was only one of many Mediterranean crops, a dominant crop in only a few localities, and there was much else in the clash of empires and religions around the shores of this sea to hold the attention of historians; but historical geographers of the sugar industry have a different perspective and for them its Mediterranean phase is extremely significant. One reason for this is environmental: the Mediterranean is the most northerly part of the world

32   *The sugar cane industry*

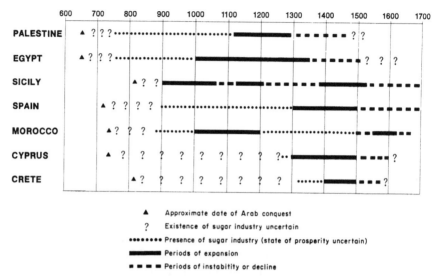

Fig 3.1 Major sugar-producing regions of the Mediterranean, 700–1700

in which sugar cane has been successfully cultivated and hence the industry here provides an early example of the adaptation of its cultivation to marginal environmental conditions. Of perhaps even greater interest is the fact that it was around the Mediterranean that cultivation of sugar cane became part of western agricultural experience. The organization of the Mediterranean industry, as it evolved during the fourteenth and fifteenth centuries, heralded the organization of the Atlantic and American colonial industries. Indeed, the Mediterranean sugar industry can be seen as a school for the colonizers of Madeira, the Canaries and tropical America. It is therefore an important link in the chain of diffusion and development that has taken sugar cane from indigenous garden plant in New Guinea to agro-industry in the tropical world of today.

The most important sources for the study of this phase of the sugar industry are the accounts of the cultivation of the plant by contemporary travelers and agronomists, both Christian and Muslim, the correspondence of merchants and the evidence of taxes, duties and prices in customs house records and other government documents. The archives of the Mediterranean lands have been slow to yield information on such important matters as the area of land cultivated in sugar cane, crop yields, manufacturing techniques and production costs, and without reliable statistical data of this type comparisons of production over time and between different parts of the Mediterranean are difficult. Recently, archaeology has provided new evidence of milling technology and manufacturing techniques (Berthier 1966a and b; Wartburg 1983). Drawing on these varied sources, it is possible to outline the rise and decline of the Mediterranean industry and to describe the techniques of cultivation and manufacture.

Establishment of sugar cane cultivation

The development of the sugar cane industry in the Mediterranean was part of the westward diffusion of the agricultural revolution begun in the Middle East under the aegis of the Arabs in the years following the founding of Islam. In the islands and peninsulas of southern Europe and along the coast of North Africa, this revolution was characterized by the cultivation of many tropical and subtropical crops brought from the East, and, through the widespread introduction of irrigation, it led to an intensification of land use because the hot summer months could now be used for cultivation. Some of the crops and techniques associated with the revolution had already reached the Mediterranean from the Near East by classical times, but, as in the Middle East, one role of the Arabs was to accelerate the adoption of the crops and agricultural techniques. Southern Spain emerged as a major locus of this new agriculture, and in time it became a center of diffusion back to North Africa and on to the New World (Watson 1974, 1983; Glick 1979: 175–97).

The first references to the cultivation of sugar cane around the Mediterranean come from Syria, Palestine and Egypt in the years after their occupation by the Arabs in the first half of the seventh century. Cultivation spread through the valley and delta of the Nile, along the Levant coast wherever there was water for irrigation (Deerr 1949–50: vol. 1, 74–87). In the second half of the seventh century the Arabs swept across North Africa, reaching Morocco in 682. Deerr (1949–50: vol. 1, 79–80) and von Lippmann (1929: 239–40) report the arrival of sugar cane in the western Mediterranean within a few years of the conquest, in accord with the adage that sugar followed the Koran. Although the plant may have traveled quickly, virtually in the baggage-trains of the armies, an industry that was dependent on irrigation, technology and reasonably peaceful times developed slowly. Two centuries or so appear to have elapsed between the Arab conquests in the western Mediterranean and the emergence there of a commercial sugar cane industry.

In North Africa, the first reference to production comes from Morocco and occurs in the work of Abu Hanifa, an author who died in 895 (Berthier 1966a: vol. 1, 43). Reports by Ibn Hawqal, a tenth-century writer, confirm the presence of an industry in North Africa. By the eleventh century, sugar cane production existed around Gabès and Djalula in Tunisia and around Ceuta in Morocco; the most significant area of production was in southern Morocco in the Sous and neighboring valleys on the flanks of the High

Atlas (Vanacker 1973: 677). Spain's first account of a sugar cane industry is in the so-called Calendar of Cordoba, which listed the major activities of the agricultural year and dates from 961, two and a half centuries after the Arabs crossed to Spain. The beginning of the industry should be placed some years earlier than the Calendar, at the opening of the tenth century or even in the late ninth century (Dozy and Pellat 1961; Imamuddin 1963: 116). The Mediterranean coast of Andalucia and the Guadalquivir valley were the main centers of the industry in Spain, although sugar cane was grown as far north as Valencia. The Arabs first invaded Sicily in 655, but they did not achieve full mastery of the island until 877. There is a record of the export of sugar from Sicily about 900, and Ibn Hawqal described the industry as flourishing half a century later (Deerr 1949–50: vol. 1, 76).

The Norman conquest of Sicily in the eleventh century and the Crusades brought northern Europeans into greater contact with the cane-producing lands. The increased familiarity with sugar among Europeans led to a growth in demand that stimulated an expansion in cultivation in Palestine and the development of industries in Rhodes, Malta, Crete and Cyprus. The most northerly extension of cultivation occurred in the last years of the industry and was experimental in nature. Early in the fifteenth century the Genoese, encouraged by the Portuguese crown, attempted to establish an industry in the Algarve. In the 1450s cane was reported growing as far north as Coimbra (Parreira 1952: 18–19). Tuscany in the 1550s was the scene of a short-lived experiment, while in the 1560s and 1570s Catherine de Medici tried to cultivate sugar cane in her gardens at Hyères in Provence (Deerr 1949–50: vol. 1, 79, 85; Jones 1966: 370).

In the thousand year history of the industry, there probably were few Mediterranean coastal valleys or plains with water for irrigation where sugar cane was not cultivated. Our knowledge of the existence of minor areas of cultivation depends on the chance survival of documents. For example, we know that cane was cultivated for a while at least in the Greek Morea, in southern Italy (Deerr 1949–50: vol. 1, 79, 83) and also in Turkey (Cahen 1968: 158) but we do not know exactly where along the extensive coastlines nor how important the crop was to the local economies (Fig. 3.2).

Organization of the sugar industry

The organization of resources for the sugar industry varied around the Mediterranean and changed through time. Uniformity was greatest in the manner of cultivation, where the environment acted as a unifying force, and in the manufacturing of sugar; diversity was most pronounced in the type of labor and in the organization of landholdings.

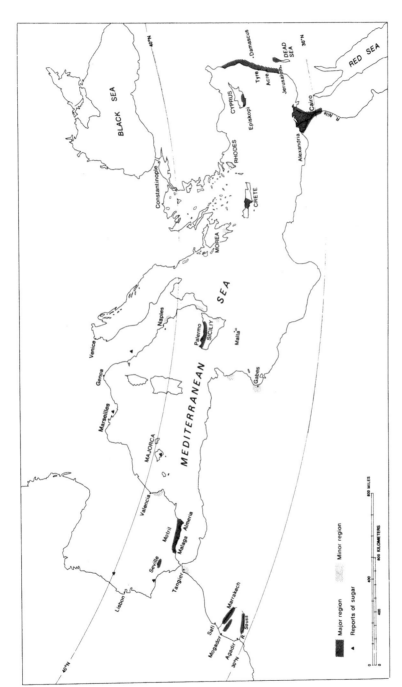

Fig. 3.2 Sugar in the Mediterranean lands, 700–1600

Cultivation

The Mediterranean is a marginal location for the sugar cane industry. The long summer drought limits cultivation to areas with abundant water for irrigation, and cool winter weather inhibits growth of sugar cane. Through much of the region frost is a hazard and its frequent occurrence prevented commercial cultivation along the northern coasts. The climate meant the Mediterranean industry depended on what was by tropical standards an immature cane with a low sugar content, a circumstance that placed the industry at a severe disadvantage when competition from New World producers began. The sugar cane was planted in February and March and harvested in January (Dozy and Pellat 1961: 36, 60, 144, 172; Renaud 1948: 30). Despite the risk of frost, the medieval Mediterranean cultivators followed the practice of ratooning, and a comment by Ibn al-Awwam (1802: vol. 1, 392) suggests that in southern Spain three harvests were cut from the same roots: the plant cane and two ratoon crops. Contemporary accounts suggest that experimentation in sugar cane cultivation was part of the Arab agricultural revolution. The Arab cultivators had tried different methods of planting cane and made recommendations on the spacing of setts in the fields and on the depth of the furrows. They knew that to bring on the germination of setts in preparation for the short growing season, the setts should be stored lightly covered with soil in readiness for planting, a practice still recommended today to cultivators in various parts of the world (Barnes 1974: 270). They had investigated closely the importance of manuring. In Spain, the manure of sheep, mules, asses, doves and other animals and birds was considered to have distinctive merits appropriate to different crops and soils. The manure was applied directly to the soil or placed in the irrigation channels to be spread by the water (Bolens 1972a). Understandably, a concern for irrigation pervades the Arab accounts. Water was to be applied every few days. In Spain the recommended frequency varied in part at least with the stage of growth of the cane, whereas in Egypt cane was irrigated twenty-eight times between planting and harvest (Bolens 1972b).

Despite the advances made in sugar cane cultivation through experimentation in planting, manuring and irrigation, the major handicap – the winter – could not be overcome. A description from sixteenth-century Sicily suggests that an attempt was made to lengthen the growing season by starting cane in protected nursery beds. Setts were placed in compost and after germination transplanted to the fields. These setts were cut from the top of the cane stems, for the buds higher on the stem germinate more quickly than those lower down (Deerr 1949–50: vol. 1, 78–9). Ibn al-Awwam (1802: vol. 1, 390) wrote of "planting cane from its roots," an imprecise phrase that could refer to ratooning but might also be an allusion to the transplanting of young cane from nurseries. The breeding of cane to produce new varieties with greater tolerance for cold weather or with the ability to mature more quickly was beyond the scope of the medieval agronomists. The discovery that cane can carry fertile seed and the beginning of selective breeding date only from the nineteenth century. Even so, the medieval agronomists advised cultivators to select the setts carefully, and over the centuries this selection may have resulted in the gradual improvement in the quality of cane.

The early years of the Mediterranean industry probably witnessed a greater measure of experimentation in sugar cane cultivation than any subsequent period in its history until the nineteenth century. Unfortunately, the surviving records do not provide a measure of the improvement in yields that this attention to the techniques of cultivation achieved.

Milling and manufacture

In contrast to experimentation in cultivation, the milling and manufacture of sugar in the medieval Mediterranean changed little. Thus the region was aptly characterized as one of "technological retardation" compared with northern Europe (White 1962: 88). The mills and presses used to extract the juice from the cane were adapted from those already used around the Mediterranean to mill flour, to extract oil from olives or to crush grapes and other fruits. The work of milling was labor-intensive and, given the nature of the equipment, inefficient (Feldhaus 1954). The cane stems were chopped into pieces several inches long and the juice extracted in two stages: the cane was first milled, which released some of the juice, and then the crushed residue, sometimes packed into sacks, was transferred to a press by which more juice was released.

The mills were generally of two types. One consisted of an upper rotating grindstone set over a lower immobile one: the pieces of cane were ground between the two stones. The second, known in the literature as the edge-runner, consisted of a wheel-shaped grindstone set upright in a shallow depression in which the cane was placed and a driveshaft protruded horizontally from the center of the wheel out beyond the rim of the depression. Men or animals turned the wheel around the depression, thereby crushing the cane. The edge-runner was widely used around the Mediterranean to crush olives, nuts and even mineral ores and had a long life in the sugar industry, for it was taken to the New World where it continued in use until the early seventeenth century (Plate 4). The final extraction of the juice was carried out by either beam or screw presses. A beam press consisted of a large heavy wooden beam or tree trunk to which was attached a smaller beam with a hammer-head: the main beam was winched or screwed down so that the hammer-head pressed on the cane placed beneath

38 The sugar cane industry

Plate 3 A Sicilian sugar mill of the second half of the sixteenth century
Source: Straet 1600: plate XIII.

on a stone surface. A screw press consisted of a series of boards, one atop the other. Cane was inserted between the boards, and the whole tightened by the turning of a screw or screws. Other simpler and even less effective methods of extracting juice existed, such as the mortar and pestle described by Crusaders in Palestine (Baxa and Bruhns 1967: 9). The most detailed description of milling in the medieval Mediterranean is Al-Nuwairi's account of work in an Egyptian mill (Deerr 1949–50: vol. 2, 536; Chapoutet-Remadi 1974: 32–4), and an engraving survives of a sugar factory in Sicily *c.* 1570 (Plate 3). In the foreground of this engraving there is a water mill; in the rear, a screw press.

The few instances of technological innovation in sugar milling during the Mediterranean era are poorly documented and it is difficult to know how they affected the efficiency of the industry. E. Ashtor (1981: 105–6) considers that the substitution of horses for oxen in the mills in the Christian countries led to improved efficiency which helped to give the western Mediterranean a competitive edge over the Levant. The substitution was gradual: Ashtor does not date its beginning but considers that it was complete by about 1400. His conclusion may be too sweeping because horses did not entirely replace oxen as work animals around the Mediterranean, and oxen continued to be an important source of power in the sugar mills of the New World. The application of water power to sugar milling was much more significant than the substitution of horses for oxen. The use of water power in the sugar mills became widespread, particularly in Palestine, Egypt, Morocco and Sicily. It was used in sugar mills built by the Crusaders, though it is not known whether they were the first to use water power for milling sugar in Palestine. The ruins of two of these mills still stand, one near Jericho, the other near a-Safi (Benvenisti 1970: 254). Al-Nuwairi mentions water-powered sugar mills in Egypt. Berthier (1966a: vol. 1, 111, 133–44, and maps) identified fourteen sugar factories in southern Morocco that employed water power. He has also traced the routes of the aqueducts from the Atlas Mountains, but was unable to date the construction of the mills. A well-designed water mill made the crushing of sugar cane more efficient, but even the use of water power did not necessarily eliminate the second stage of extraction in the presses, as the mill shown in Plate 3 demonstrates. The Mediterranean industry has been credited with making an advance of fundamental importance in sugar mill technology, the invention of the three-roller mill. Two notable historians of the sugar industry (von Lippmann 1929: 338; Deerr 1949–50: vol. 1, 7; vol. 2, 535) as well as some more recent scholars (Baxa and Bruhns 1967: 16; Barrett 1970: 53; Verlinden 1970: 20) attribute the invention to a Pietro Speciale in Sicily in 1449. This attribution is almost certainly incorrect (Galloway 1977: 186–7), and this new design of mill was not used by Mediterranean millers of sugar. The first undoubted reports and illustration of the new mill in the western world come from Peru and Brazil, c. 1610, and its likely place of origin is China (see ch. 4, pp. 73–5, ch. 8, pp. 197, 206).

The manufacture of sugar from the juice revealed a further limitation of the Mediterranean environment – the shortage of fuel. Even by the time of the Arab conquests, the Mediterranean forest had already been heavily depleted. Deforestation, with the resulting shortage of timber, was particularly acute in the southern Muslim lands. The progress of industries, such as metallurgy and the manufacture of pottery, glass and sugar, that required fuel was severely curtailed (Lombard 1957, 1959), and Berthier (1966a; vol. 1, 81), for example, has attributed the absence of sugar cane cultivation from some of the irrigable valleys in Morocco to the lack of timber for fuel. No evidence has emerged of the use of cane bagasse – the residue of the sugar cane after milling – for fuel around the Mediterranean as it was to be used later in tropical America when wood became scarce. In fact, the use of bagasse for fuel depended on improvements in mill and furnace design that only came in the seventeenth century. The scarcity and expense of fuel was one reason why the Mediterranean industry remained small-scale and sugar during these centuries a luxury item.

The Mediterranean refiners, like their Indian forebears, made various qualities of sugar. The basic process was to boil and concentrate the juice, skimming off impurities as they rose to the surface. At the point of crystallization, the syrup was poured into inverted earthenware cones in which it cooled, crystallized and packed into the traditional sugarloaf shape. Molasses drained from the crystallizing sugar through a hole in the tip of the cone and could be reboiled to produce more sugar. The finest quality sugar was made by dissolving in water the crystals from the first boiling and reboiling and recrystallizing two or three times. Sugar was marketed in a variety of grades, in powder, in lumps and in loaves. Some was colored and flavored with violets or rose water (Pegolotti 1936: 362–5, 434–5). In Morocco (Berthier 1966a: vol. 1, 193–201), and quite probably elsewhere, sugar was "clayed" by placing waterlogged clay over the top of the cones. The water from the clay percolated through the sugar, leaching out the lingering traces of molasses and made a sugar loaf that was whitest near the clay and graded into dark brown at the base. The manufacture of "clayed" sugar was later to become widespread in the Americas.

Towards the end of the fifteenth century, there were major changes in the geography of sugar refining. Because sugar cane is perishable and bulky, milling and the manufacture of at least a crude crystal sugar had to take place among the cane fields after which, theoretically at least, the further refining of the sugar and export of a finished product was optional. After about 1470, European importers began to build refineries, first in Venice and Bologna, then in Antwerp, and during the sixteenth and seventeenth centuries in many northern cities, where they converted crude crystal sugar into superior grades. This transfer of part of the manufacturing process from producing country to importing country had both causes and consequences. Much sugar arrived water-damaged after the sea voyage, and it therefore made good sense to conclude the refining after the risks of transportation were passed. Fuel, moreover, was more readily available and cheaper in the north than where the cane was grown. The fact that the transfer moved employment in the industry from producer to importer may or may not have been a consideration in the minds of the importers. Naturally, the development of refining in the north reduced the producer's interest in making fine quality sugar and, most significantly, had the effect of making the producer subservient to the importer. In brief, whatever the motives of the various investors in refineries, in addition to their desire to make money, the relocation of sugar refining to the importing countries placed the producer in a dependent or "colonial" relationship with the manufacturer, a relationship that has survived with little change to this day.

Land and labor

The pattern of land tenure and the type of labor employed in the sugar industry varied greatly from one part of the Mediterranean Basin to another, and changed with the passage of time. During the early years of the industry, the organization of land and labor was similar in the Muslim lands of Spain, North Africa and the Levant. Later, developments in Egypt and Morocco broke this uniformity; the situation was also different in the Christian islands of Crete and Cyprus.

In the Muslim countries the size of landholdings ranged widely. Peasant-owned plots existed side by side with large estates. The estates for the most part were parceled out to tenants who farmed the land as sharecroppers. The percentage of the harvest that the tenants had to surrender to the owners depended on the quality of the land and on the proportion of the total investment in seed and equipment made by the landowners and tenants. Only a small amount of land was held as demesne, land that the owners undertook to cultivate by drawing on the labor of a serf class. Indeed, corvée and slave labor were rare; the agricultural work was done by tenants, their families and even hired hands. The question of whether, in this mix of landholdings, sugar cane was exclusively an estate crop or was also grown on peasant holdings remains obscure (Watson 1974: 29–30). In Spain this pattern of rural organization appears to have been long-lasting (Imamuddin 1963: 60–4; Lévi-Provencal 1967: vol. 2, 267–8) and it survived in the Levant until the end of the Crusades (Cahen 1940). In the Crusader states, the land passed into the ownership of western feudal nobles, of military orders such as the Knights of St. John, of the church and of Italian merchant cities (Riley-Smith 1967; Prawer 1952–3, 1972; Williams 1974: 73), but continuity with Muslim agricultural tradition was maintained. Sugar cane, however, perhaps because of the large capital investment it required, was grown on the demesne land and a corvée was levied on the peasantry to provide the labor (Riley-Smith 1973: 46; Prawer 1953: 165).

The history of land tenure and agricultural labor in both Egypt and Morocco is difficult to unravel, but it is clear that the approach to land management differed from that in other Muslim countries. More information is available on Egypt than on Morocco. In Egypt, in Mamluk times (1250–1517), the land was controlled by the state – or, more precisely, by the sultan – and divided into estates known as *iqta's* which the sultan awarded to Mamluk military officers. The *iqta's* were not hereditary, and their revenue was gauged to the rank of the officer. The officers usually lived in Cairo, leaving the management of the *iqta's* to agents. Cultivation was carried out by peasants who paid a tax or a portion of the crop to the grantee. Corvée in Mamluk Egypt was rare, but it was used in the cultivation of sugar cane (Poliak 1936: 262; Rabie 1972: 26–72). The record of landownership in Morocco is almost completely lost. Berthier (1966a: 224–9, 240–2, 1966b: 37, 40) concludes that the sugar industry was a state monopoly and that the state farmed out the management of the mills and estates. The record is also incomplete on the type of labor employed. He also accepts largely on the basis of place name evidence – names of several locations in the valleys in which sugar cane was cultivated incorporate the word "slave" – that the sugar industry used slaves. Certainly, the development of the slave trade was one of the motives for the Moroccan trans-Saharan expeditions of the late sixteenth century.

Estates on Crete and Cyprus differed from those in Muslim countries in that demesne land was much more extensive and the corvée was an important source of labor (Riley-Smith 1967: 105). During the fourteenth and fifteenth centuries, agricultural labor on these islands became scarce because of the ravages of war and plague; in response to this shortage slave labor was increasingly used. Even before the Black Death in 1348, slaves were being imported to Crete and Cyprus, but later slavery became even more significant. The slaves were from varied national backgrounds: Greeks, Bulgarians, Turkish prisoners of war and Tartars brought from the shores of the Black Sea (Thiriet 1967; Verlinden 1970: 26–32).

As the organization of the Mediterranean industry evolved, the antecedents of plantation agriculture can be recognized. The cultivation of sugar cane in many parts of the Mediterranean employed forced labor, at first corvée and later also slave labor. The link between sugar cane cultivation and slavery which was to last until the nineteenth century became firmly forged in Crete, Cyprus and Morocco. In addition to forced labor, there were other harbingers of plantation agriculture in the Christian-ruled lands of the eastern Mediterranean. For example, a colonial relationship was established between the primary producing, cane-growing areas and the metropolitan, manufacturing and refining centers of Europe. When Crete and Cyprus passed under Venetian rule, in 1204 and 1489 respectively, they became colonies in both political and economic senses of the term. Venice encouraged the agricultural development of its colonies in the eastern Mediterranean and looked to them for supplies of wheat, wine, raisins and other products in addition to sugar (Thiriet 1959). In the fourteenth and fifteenth centuries sugar increased in importance, especially in Cyprus, and large estates were devoted to it. One example is the estate of the Cornaro family at Episkopi which in the mid- and later 1400s was reported to be employing 400 laborers in the production of sugar (Hill 1948: vol. 2, 816; Verlinden 1970: 19–20). The Cornaros were of Venetian origin and the last Queen of Cyprus was a member of the family: given this fame and wealth, the Episkopi estate was probably atypically large, but nevertheless it indicates that at least one Mediterranean sugar estate was comparable in terms of numbers of workers to the largest plantations of colonial tropical America.

The last phase: 1300–1600

The decline of the Mediterranean sugar industry has traditionally been attributed to competition from more efficient producers in the new European colonies in the Atlantic and America. Madeiran sugar began to reach Europe after 1450, and by 1500 was being distributed throughout western Europe, finding markets even as far east as Constantinople and Chios, the Genoese colony and entrepôt off the coast of Asia Minor (Rau and Macedo 1962: 12–16; Heers 1961: 495–7). Sugar from São Tomé first appeared in Europe during the 1490s and Brazilian sugar began to arrive in the 1530s and 1540s (Tenreiro 1961: 67–74). This traditional explanation of the decline of the industry appears overly simplified when set against the changing geography of sugar cane cultivation during the last phase of the Mediterranean industry. The production of sugar decreased in importance in Egypt, Palestine and Syria during the fourteenth century; there was a subsequent increase in production in Cyprus, Crete and the western Mediterranean. The final collapse of the industry throughout the Mediterranean did not come until the late sixteenth century. In other words, the decline of the sugar industry in the eastern Mediterranean began more than a century before, and the industry in the west flourished for more than a century after Madeiran sugar first appeared on the scene. To account for the early decline of the industry in the East factors other than competition must be considered: they are warfare, plague, the policies of the Mamluk sultans of Egypt and technological stagnation.

Egypt, Palestine and Syria, the three countries that had been exporters of sugar to western Europe, were, in a reversal of this early trading pattern, by the end of the fifteenth century importing sugar from the West (Ashtor 1969: 384). The decrease in the number of sugar refineries in Egypt is indicative of the change: there were sixty-six refineries in Fustat (Old Cairo) in 1324; a century later only nineteen were functioning, the others having been abandoned or converted to new uses (Udovitch 1970: 116). Other forms of agriculture suffered along with sugar cane cultivation. Debate continues over the cause of this prolonged decline in agriculture and prosperity, but warfare is a readily available explanation. The wars that finally overcame the Crusader states in Syria and Palestine during the second half of the thirteenth century were followed by the Mongol invasions that culminated in the ravages of Tamerlane in 1400, and then, in the fifteenth century, the expansion of the Ottoman Turks led to yet another round of conflict. Inevitably, these wars in the eastern Mediterranean were destructive of agriculture and trade.

However, warfare was not the sole or even the most significant cause of the decline in agriculture. In Egypt, Mamluk misrule led to a downward spiral in the economy. The weakening of government authority in the countryside opened the way for the bedouin of the desert to plunder the villages with the result that peasants sought refuge in the cities and left their land untended. These developments in turn resulted in a fall in the revenues of the *iqta's*. The Mamluks sought to remedy this loss in their income by raising taxes on commerce and by establishing monopolies in the most profitable trades, including that of sugar. Inefficient and corrupt administration of the monopolies further impeded agriculture. Al-Makrisi, a contemporary witness, appears to have been the originator of this explanation of the decline of Egyptian agriculture and he has been followed by some modern scholars (Ashtor 1969, 1981; Darrag 1961). Recently, this interpretation of Mamluk Egypt has been questioned. The revisionist explanation reverses the line of argument: the Mamluk policies were not the cause of the decline, but a response to depopulation brought about by plague (Dols 1977; Udovitch 1970).

The Black Death, which reached Egypt in the fall of 1347, was one of a series of plagues that struck there in the fourteenth and fifteenth centuries. By the early 1400s, the population of Egypt may have been reduced by as much as a third. Other factors contributed to the population loss. Marked fluctuations in the level of the Nile as well as poor maintenance of the irrigation system contributed to crop failures and famine. Evidence also points to a cattle murrain at the time of the Black Death that would have seriously reduced the stock of plow and work animals. In addition to the general decline in population, there was a migration of people from the country to the cities, for studies show a substantial reduction in the number of villages in Egypt. This movement was spurred by the privations in the countryside, by a search for medical attention during the time of plague and by the attraction of high wages paid in the cities because of a general shortage of labor. Following the Black Death, the price of goods that required a substantial labor force to manufacture greatly increased. Sugar was no exception. Beset by war and plague, during the fourteenth and fifteenth centuries Egypt ceased to be an important source of sugar.

The repercussions of plague and warfare rather than technological stagnation and Mamluk misrule led to the decline of sugar production in the Levant, but lands farther west also suffered from war and plague. However, the decline of such major producers as Egypt and Palestine gave an opportunity to western producers of sugar to expand their activities. Capital investment, of which the development of a slave trade is one example, overcame losses due to war and disease. Venice and Genoa, deprived of their lands in the Levant, actively supported the development of the industry elsewhere in the Mediterranean. The Cypriot and Cretan industries expanded in the fourteenth century and flourished in the fifteenth. An increase in production in Granada, where the Genoese played an important role in the trade, converted this last Moorish kingdom on the Iberian peninsula into a virtual colonial territory of Genoa. The Genoese also attempted to establish sugar cane cultivation in the Algarve and became heavily involved in the industry in the Canaries and Madeira (Heers 1957, 1961). In Italy, there is evidence of a redirection of capital investment from commerce to agriculture (Lopez 1970: 107–15; Jones 1966: 370). The expansion of the western Mediterranean industry continued throughout this period even though sugar had begun to arrive from Madeira and the Canaries. This increase in the supply led to a fall in the price of sugar, but the new producers did not replace the old, for in the growing economy of the "long sixteenth century" Europe was able to absorb the sugar from the Atlantic colonies and from the Mediterranean.

The end of the sugar industry in Cyprus, Crete and the western Mediterranean came in the brief period of thirty years, approximately 1570 to 1600, with in many instances cotton taking over as the new cash crop (Hill 1948: vol. 2, 817; Masefield 1967: 290; Trasselli 1957: 146). By the end of the century, the Granadan industry had been reduced to seven mills and a few acres of cane in the vega of Motril (Blume 1958: 98–9). After 1600, the cultivation of sugar cane was no longer of commercial significance around the Mediterranean except along the southern coast of Spain but it survived as a garden curiosity in Valencia, Sicily and elsewhere into the eighteenth century.

The immediate cause of the collapse of the industry in the western Mediterranean was the arrival in Europe of Brazilian sugar at a price below that at which sugar could profitably be produced in Mediterranean industries. In 1580, Brazil and Portugal came under Spanish control; therefore, Brazilian sugar could enter Spanish Mediterranean dominions, and it undersold Sicilian sugar even in Palermo (Trasselli 1957: 146). Brazilian sugar was handicapped by high transportation costs but enjoyed environmental advantages. Sugar cane could grow to maturity and yield more sugar in Brazil than around the Mediterranean. It was cultivated without the expense of irrigation and there was an abundance of fuel and land. Around the Mediterranean, land, especially irrigated land, was scarce and valuable, and sugar had to compete for space with other crops. As imports of inexpensive Brazilian sugar increased, the time came in the late sixteenth century when other Mediterranean crops were more profitable to grow than sugar cane.

In addition to Brazilian competition, other factors contributed to the difficulties of the industry in the western Mediterranean. In Morocco, the decline of the sugar industry coincided with a long period of civil strife which was adversely affecting production, according to sugar traders, by 1576. Over the next decades there were reports of damage to mills, but statistics on production do not survive. Finally, in 1622, Antony Sherley, an English merchant in Granada, reported that the Moroccan industry no longer existed (Berthier 1966a: vol. 1, 269–72). Catholic prejudice against the Morisco population of southern Spain may have deprived the industry of skilled workers. The Moriscos were first dispersed from Granada after the Christian reconquest in 1492 and then many of their descendants were expelled from Spain between 1609 and 1614. The significance of the expulsion may have been exaggerated because research has shown that it was not as complete nor the economic consequences for Spain as grave as was once thought. Moreover, many of those expelled from Spain came from Valencia, which was not a major center of sugar production (Lapeyre 1959; Braudel 1972–3: vol. 2, 792). No evidence exists that the fuel supply dramatically worsened in the second half of the sixteenth century, but nevertheless deforestation added to the difficulties of the industry. The progressive destruction of the Mediterranean forests not only made fuel increasingly scarce and presumably more expensive but also caused soil erosion in the hills and silting in the plains, which further complicated the maintenance of irrigation.

The rise and fall of the Mediterranean sugar industry has also been linked to climatic change because of the rough coincidence between its establishment in a notably warm period, which reached a peak about 1000–1200, and its decline in the western areas with the onset of the so-called Little Ice Age of approximately 1550–1700. During the warm or optimum period, temperatures were about 1–2 °C above present values to the north of 40 °N and rather less to the south. In the south, during the warm period there was probably much more rainfall than at present (Lamb 1965). It is difficult to assess the impact of such minor changes on sugar cane cultivation. Presumably the risk of frost damage during the Mediterranean winter did not disappear during the warm period, though it must have been less in the climatic optimum than in the later cooler period. Le Roy Ladurie (1971), using examples from northern Europe, recently cautioned against accepting climatic rather than economic explanations for agricultural change. To accept climatic fluctuations as the significant force in the rise and decline of the Mediterranean sugar industry decreases the role of economics and major historical events. Moreover, the rapid collapse of the Mediterranean sugar industry, in a thirty-year period, suggests that climatic change was not a major cause, for it is a gradual process that makes its effects felt only over many decades, if not centuries. Competition from American sugar rather than environmental change brought the Mediterranean sugar industry to an end.

The first phase of the Mediterranean sugar industry was marked by experimentation. After the Arab agricultural revolution, agricultural techniques changed little, if at all, and even milling remained largely unchanged with the continuing use of technology dating from classical times. Water power was the only major innovation in milling, although it did not supplement other forms of power. During the later years of the industry, change was most obvious in the organization of trade, labor and land. Gradually, the characteristics of the colonial plantation system emerged. Aspects of the system were discernible even in the Crusader states and became more noticeable in the development of the sugar industry in Crete and Cyprus. The settlement of Madeira, the Azores, the Canaries, and São Tomé marked further stages in the development of the new system, so that by the early sixteenth century, when the settlement of Brazil began, the plantation system was already a tested form of colonial land use. In the Americas, with an abundance of land, an ideal climate and a supply of slave labor, sugar production found scope to flourish.

4

The Atlantic sugar industry: *c.* 1450–1680

The title of this chapter derives its justification from the new Atlantic distribution of the sugar industry (Fig. 4.1). Portuguese and Spanish discoverers carried sugar cane with them to the colonies they founded on both sides of the Atlantic Ocean in response to the continuing demand for sugar that turned the industry into an instrument of European imperialism, a means both of financing colonial endeavors as well as a motive for the occupation of yet more territory. Another equally apt title for the chapter would be "The Spanish and Portuguese sugar industry," because these were the years during which the two nations held a monopoly on the supply of sugar to Europe. Eventually, production of sugar in the earliest Atlantic colonies began to decline in face of the competition from Brazil so that by the time the English, French and Dutch broke the Iberian monopoly by founding their own plantations in the second half of the seventeenth century the sugar industry had become almost entirely American in its location.

In this migration from the Mediterranean to America, the industry experienced very considerable modifications. Madeira, the Azores and the other islands off the West African coast have long been regarded by historians as training grounds where the Spanish and Portuguese learnt lessons in colonial administration they later applied in the Americas. So it was also with the sugar industry. On these islands the colonists learnt to adapt their Mediterranean techniques of cultivation and production to cope both with unfamiliar environments and the ever larger scale of the industry, processes in which those initial characteristics of plantation agriculture first discernible in the eastern Mediterranean became much more pronounced. They soon realized that sugar cane would not grow well on all of their Atlantic colonies. The climate of Lanzarote and Fuerteventura in the Canaries was too dry and without water for irrigation. The Cape Verde Islands also had the drawback of having too little water, and their economy came to depend on the export of salt and the supply of goat meat to passing ships rather

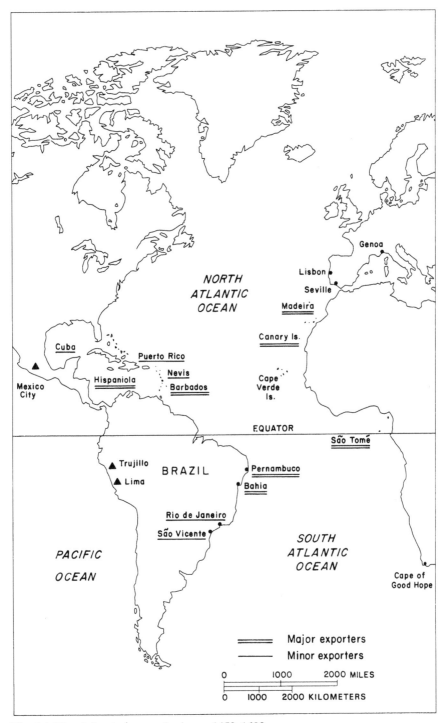

Fig. 4.1 The Atlantic sugar colonies, c. 1450–1680

than on sugar. Sugar cane was cultivated on the Azores, but the relatively cool temperatures made for disappointing crops. On Madeira, on the Canarian islands of Grand Canary, Gomera, La Palma and Tenerife as well as on São Tomé in the Gulf of Guinea conditions for the cultivation of sugar cane were considerably better than around the Mediterranean and these colonies have an important place in the historical geography of the sugar industry. Within a few years of the discovery of America, the Spanish and Portuguese began the commercial cultivation of sugar cane on Hispaniola and along the coast of Brazil, where at last they encountered an ideal climate for the crop and also seemingly unlimited supplies of land and fuel. The great production of the American colonies made sugar cheaper than it ever had been before with the result that it came within reach of a larger group of consumers, moving down so to speak below the salt: by the later 1500s, sugar, whether white or brown, in loaves, lumps or powder, appeared not only on the tables of princes but on those of the bourgeoisie too.

The Atlantic islands

Madeira

The date of the discovery of Madeira remains uncertain. It may have been sighted from Genoese or Catalan galleys making rare, bold forays out along the Atlantic coast of Morocco, but these voyages led neither to permanent settlement nor to claims of sovereignty. Portuguese interest in what lay over the horizon of the open seas to the south and west of the homeland apparently only stirred after their conquest of Cueta in 1415. They found in Madeira a high, steep-sloped island of volcanic origin, well-forested, uninhabited, with some 20 miles off shore a much smaller, drier island now known as Porto Santo. The first settlers arrived in 1425. The forests sustained for some brief years a trade in timber, but the colonists soon turned to the cultivation of wheat to help redress Portugal's chronic shortage of grain (Serrão 1954: 340). This economy was disrupted by the introduction of sugar cane.

The earliest reference to sugar on Madeira dates from 1433 and twenty years later it was being produced in sufficient quantities to initiate a modest export to Portugal, Flanders and England (Rau and Macedo 1962: 4). The explanation for the slow progress probably lies in the effort required to terrace the steep hillsides and to cut the ditches now known as *levadas* to carry rain water from the upper reaches of the hills to the fields of sugar cane. Sometime during the second half of the 1400s sugar cane replaced wheat as the principal crop on the island. The record of the export of sugar, although far from complete (Fig. 4.2) provides the best guide

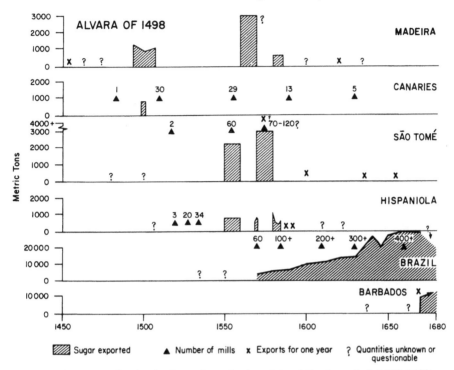

Fig. 4.2 Sugar production in the major colonies of the Atlantic period, 1450–1680

Sources: Barbados: Dunn 1969: 4. Brazil: Lockhart and Schwartz 1983: 249. Canaries: Fernandez-Armesto 1982: 81; Matznetter 1958: 91–3. Hispaniola: Chaunu and Chaunu 1956–9: vol. 6, 1008. Madeira and São Tomé: Mauro 1960: 184–92; Rau 1964: 5; Silva 1958: 83–5.

to the growth of the industry. Madeira was exporting small amounts of sugar during the 1450s and 1460s, but thirty years later annual production had grown to 100,000 *arrôbas* (1,468 tons), nearly all of which was exported. In an *alvará*, or charter, of 1498, the crown set a ceiling on the annual export of sugar at 120,000 *arrôbas* (1,762 tons), a quantity that Rau and Macedo (1962: 14–15) in their discussion of the imperfectly understood reasons for introducing the regulation regard as being about the maximum the island was capable of producing. Indeed in the decade following 1498, exports did not nearly reach this ceiling. The framers of the regulation anticipated a distribution of the exports in the manner shown in Table 4.1, and although this table represents a plan rather than fact, it was presumably based on existing trade flows and so serves to demonstrate how deeply by this comparatively early date Madeiran sugar had penetrated Mediterranean markets. After 1500, with the expense and labor of clearing the land and of building an irrigation system well in the past, Madeira

Table 4.1 *Sugar exports of Madeira as anticipated in the* alvará *of 1498*

	Arrôbas	Tons
Portugal	7,000	103
Flanders	40,000	587
England	7,000	103
Rouen	6,000	88
La Rochelle	2,000	29
Brittany	1,000	15
Aigues Mortes	6,000	88
Genoa	13,000	191
Leghorn	6,000	88
Rome	2,000	29
Venice	15,000	220
Chios and Constantinople	15,000	220
Total	120,000	1,762

Source: Rau and Macedo 1962: 14.

entered its heyday as a sugar colony. It had become not only an important supplier of sugar to Europe but, if we are to accept the testimony of the English doctor, Thomas Muffett (1655; David 1977: 139), its sugar was the finest on the market.

Surviving records do not provide the details of the Madeiran industry that we would like: discussion of the agricultural routine, of the organization of labor and of the finances has not come down to us. We do know that most of the sugar – perhaps 75% of the total – was grown on the lower slopes of the hills along the south coast, to the east and west of the capital city of Funchal (Rau and Macedo 1962: 14). The north coast was a less favored locale in part because of the cooler temperatures but also because of the lack of accessible beaches from which to embark the sugar. A 1494 inventory of the cane growers in the captaincy of Funchal does provide an insight into the structure of the industry at a time when it had become well-established. The inventory lists 221 owners of cane fields. There was a considerable range in the size of holdings: the 10 largest landowners produced 25% of the crop, the 50 largest 70% while 171 growers accounted for the remaining 30%. Yet, even the most extensive of these holdings were small in comparison to those that were to develop in the Americas. Only three of the owners could grow enough cane to produce more than 2,000 *arrôbas* (29 tons) of sugar; eighty-eight of the owners produced cane for less than 100 *arrôbas* (1.5 tons) each. The inventory also makes plain that members of various strata of society participated in the industry.

Fidalgos or noblemen and officials were among the cane growers as were surgeons and millers, bankers and merchants, cobblers and carpenters (Rau and Macedo 1962; Rau 1964).

If in its pattern of land tenure the sugar industry of Madeira was reminiscent of that of the Mediterranean, so it was also in the technology of its sugar mills. In this regard there is no evidence of any technological innovation having taken place on the island. In the early years, the cane was crushed in very rudimentary mills, known as *alçapremas* that consisted of two rollers one set upon the other that were turned by hand. They did not entirely disappear when later entrepreneurs with some capital behind them built larger mills powered by animals or water (Parreira 1952: 22). Prince Henry the Navigator granted the first license for the construction of a water mill in 1452, claiming as his due one third of the sugar, but this type of mill remained rare on Madeira (Mauro 1960: 186). The authorities did indeed carefully regulate and tax the milling and manufacture of sugar, levying naturally a higher revenue from the larger mills than from the *alçapremas*, but the records have not survived. The actual number of sugar mills on the island, therefore, remains in dispute, with widely varying figures appearing in the literature, particularly for the years around 1500. Von Lippmann's (1929: 403) count of one hundred and fifty does seem unrealistically high as does Mauro's (1960: 186) of eighty, particularly when set against the fact that the 1494 inventory lists only sixteen. An inference from the inventory with its evidence of numerous small-holders and few mills must be that on Madeira there was a separation between the growing of cane and the milling and manufacture of sugar. After harvesting their fields, growers took the cane to a neighboring mill where they presumably paid the miller for his work with a portion of the crop.

The role of slavery in Madeira was relatively minor. If indeed slaves were employed to clear the land and cut the irrigation ditches as has been suggested (Greenfield 1977: 541; Marques 1972: vol. 1, 158) little use was made of them subsequently in cultivating the fields. Few holdings on Madeira were large enough to require a numerous labor force and the small-holders took care of their own small fields with the help of their families. Rau (1964: 6) accepts that there were some slaves, but is at pains to point out what she calls the "restricted character of slavery" on the island. Some figures from the mid-1500s give a clearer idea of the situation. In 1552, the King permitted Madeiran cane growers to send a ship every two years to Guinea to collect slaves. In Funchal, 3,000 slaves were reported that year. In 1567, the crown authorized the import of 150 slaves a year for five years from the markets in the Cape Verdes (Mauro 1960: 185). These data suggest only a very modest trade in comparison to the traffic that was to supply later sugar colonies: Madeira continued but did not deepen the link between the sugar industry and slavery already forged around the Mediterranean. With so few slaves, Madeira was a sugar colony with a European, largely Portuguese, population. On the evidence of the family names given in the 1494 inventory, the Portuguese grew the sugar cane but the export of the sugar was in the hands of foreigners. The Genoese expanded to Madeira their interest in the sugar trade already well-developed in Granada and the western Mediterranean. Their ships plied back and forth between Funchal and Lisbon in a profitable traffic that gave one notable Genoese, Christopher Columbus, early experience of navigating the Atlantic (Pike 1966: 99, 154). In the Madeiran sugar trade, the Genoese were joined by Flemish, Jewish and English merchants.

The decline of the Madeiran sugar industry began at the same time as the decline of the industry around the Mediterranean, and for the same reason: competition from Brazilian sugar which could undercut the Madeiran price by 50% (Duncan 1972: 34). After 1580, when Brazilian sugar began to reach Europe in appreciable quantities, the decline was precipitate, so much so that within a few years exports had dwindled to the point of insignificance. The Madeirans had hoped that the government would limit Brazilian production or at least guarantee Madeira a share of the market, but this was an unrealistic expectation as the development of Brazil was of more compelling interest to the government of Portugal than the protection of a small and now redundant sugar industry. For a time, Madeirans imported the poorest quality of Brazilian sugar, further refining it to reexport at high prices for the upper end of the market. This practice kept some sugar workers active, but it did not help the cane growers. There was a brief new spurt of interest in the industry following the 1620s when the Dutch occupation of parts of northeast Brazil led to years of warfare there and to the destruction of many plantations. But the decline resumed after the expulsion of the Dutch from Recife and the return of Pernambuco to the international sugar trade, a trade that soon became even more competitive when it was joined by the new Caribbean colonies of the English and French. Madeirans were forced to change the basis of their economy once again. The sugar mills were dismantled, vineyards replaced the fields of cane and the island soon became famous in England and her colonies in North America for the sweet dessert wine that bears its name. One happy consequence of the decline of the sugar industry was the disappearance of slavery from the agricultural work force (Mauro 1960: 189). Sugar cane has survived on the island, more or less as a garden crop although there has continued to be a modest market for the local sugar in the preparation of candied fruit, confectionery and rum. In 1845, armed with the promise of a large subsidy from the government of Portugal, an Englishman, William Hinton, built a new sugar mill in Funchal (Duncan 1972: 37). There is still a small factory on a side street of the city but the patches of green cane on the terraces above the beaches are more visible reminders for today's tourists of the industry that was the mainstay of Madeira's economy over two centuries.

The Canaries

The Canary Islands were known to the Spanish for more than a century before they became sugar colonies. There is evidence, rather hazy at best, of Catalan and Mallorcan missionaries working there as early as 1352 amongst the indigenous population, a people who have become known to us as the Guanches but who now no longer exist as a distinct identifiable group (Parsons 1983: 447–8). Where there were people to convert, there were also people to exploit and slave raiding provided a second reason for interest in the islands. The islands fell to the Spanish over a period of about one hundred years, first the less attractive, later the more fertile, more densely populated and better defended islands: Lanzarote, Fuerteventura and Hierro by 1409, Gomera in 1447, Grand Canary between 1477 and 1484 and finally, at the end of the century, Tenerife and La Palma. Orchil dyes derived from the island vegetation provided the colonists with a useful export to complement the money they obtained through slaving and they soon added wheat as a cash crop. When sugar cane first reached the Canaries is not known, but the origin of the industry can be dated to the building of the first sugar mill, on Grand Canary, in 1484 (Fernandez-Armesto 1982: 14). Both the late date of the conquest of the most fertile of the Canaries so long after their discovery and the subsequent rapid introduction of the sugar industry are probably connected to the progress of the Madeiran industry which by the late 1400s had reached its full potential and could no longer be expected to satisfy the growing European demand for sugar. The time for the colonization of the Canaries came when the sugar industry needed more room for expansion.

The history of the Canaries shows the speed with which remote islands could be transformed into sugar colonies when government encouragement, capital, know-how and demand came together. Alonso de Lugo, a prominent *conquistador*, was zealous in his promotion of the industry, not only building that first mill on Grand Canary but also mills on three other islands. He insisted also that those to whom he granted land cultivate sugar cane. It was indeed a common practice among *conquistadores* to favor those who undertook to cultivate cane with larger grants of land, including irrigable land (Fernandez-Armesto 1982: 80; Hernandez 1982: 12). These settlers enjoyed the advantage of being able to use the existing irrigation system of the Guanches, a system they were over the years greatly to extend (Glick 1972: 6). Madeira supplied the original stock of sugar cane as well as many experts in all facets of the sugar industry (Hernandez 1982: 12, 25). The Genoese along with a far smaller group of Catalan, Castilian and Portuguese merchants provided the finance, ships and knowledge of the market (Camacho and Galdas 1961: 42–3). With this combination of forces working in its favor, not surprisingly the industry prospered. In 1497, Grand Canary was described as a "land of many canes." Tenerife was exporting sugar by 1506, that is within scarcely six years of its conquest. Figures on the quantities of sugar produced annually in the Canaries do not survive, but the record of the number of sugar mills on the islands indicates a rapid expansion during the first decade of the sixteenth century so that between 1510 and 1520 there were at least fifteen mills on both Grand Canary and Tenerife. These two islands were responsible for the bulk of Canarian sugar with only small quantities coming from La Palma and Gomera. The industry reached its apogee during the years between 1520 and 1550 for at no later date is there evidence of a greater number of mills (Fig. 4.2). It took little more than thirty years to transform the economy of these remote islands by introducing a highly commercial, intensive form of land use that supported a distinctive, hierarchical society.

Sources for the study of the Canarian industry are abundant in comparison to those that have survived from Madeira where the only detailed document is the 1494 inventory. Contracts between landowners, millers and laborers provide the basis of some modern studies (Fernandez-Armesto 1982; Camacho and Galdas 1961; Fabrellas 1952) and there is the contemporary description of the industry on Grand Canary that Francisco Hernandez published in Mexico City (1615: 57–9) as a manual of instruction for would-be New World cultivators. It attracted attention in Europe, being republished there in Latin translation (Marcgravius 1648: 85–6), as indeed it deserved because it is the most detailed discussion of the agricultural and manufacturing practices of the sugar industry since those of the medieval Andalucian agronomists. The major drawback in the Canarian documentation is the lack of statistics on production or exports.

In the benign climate of the Canaries, the sugar cane could grow to maturity without risk of frost damage. The first crop was ready for harvesting after two years. The roots of the cane were of course left in the ground to produce ratoons and the first ratoon, known in the Canaries as the *çoca*, was ready in a further year and a half; the second ratoon or *reçoca* could be harvested twelve months later. Normally, no more than three crops were taken, with the *reçoca* providing the nodes or seed cane for subsequent planting but if the *reçoca* had been cultivated with particular care it was worth milling for sugar. Any fourth crop was good only for seed cane. Thereafter, the fields were cleared of the old roots preparatory to replanting. The fields had to be weeded, manured, irrigated two or three times a week and cleared of such pests as rats (Hernandez 1615: 57).

This intense round of work was known as the *cura* of the cane fields.

The surviving legal documents reveal a variety of contractual arrangements between landowners and agricultural laborers who were known as *cañaveros*. In a common form of contract, a *cañavero* undertook the *cura* of a cane field in return for 10% of the produce with some supplementary provisions for expenses and supplies. The largest estates had sufficient work for fifteen to twenty *cañaveros*. In another form of contract, the *cañavero* claimed the entire crop but agreed to pay a fixed quantity of sugar to the landowner. In yet a third variant, one party to a contract provided the land, the second the water and the third the labor, agreeing to divide the cane between them. Some landowners employed their own labor, paying a salary. Presumably also, a *cañavero* who contracted to care for a large area of cane would employ his own help. The Portuguese, or more precisely, Madeirans, enjoyed a reputation as experts in sugar cane cultivation and appear frequently in the documents as *cañaveros*. Spaniards as well as Guanches also figure prominently in the labor force. In these contracts, slaves receive little mention. There were slaves on the islands, mostly black Africans with a few Berbers and possibly also some Guanches, but slaves were employed more in the mills than in the fields (Fernandez-Armesto 1982: 84–5; Camacho and Galdas 1961: 16–17; Hernandez 1982: 28–9).

These arrangements for cultivating land in the Canaries reveal the still strong Mediterranean ties of the sugar industry there as indeed does the method of milling the cane and manufacture of sugar. The mills were the *alçapremas* of Madeira consisting of two rollers one set upon the other and turned by animal power. These were so inefficient in their ability to extract juice that the cane was transferred from the mills to presses described by Hernandez as like those used in Spain to crush grapes for wine. His account of the boiling and clarifying of the sugar in a succession of cauldrons to the point of crystallization and the drying of the sugar in clay molds is in fact familiar from Stradanus' print of sugar manufacture on Sicily.

The mill, press and factory not only represented a large investment of capital but carried heavy operating costs. The work required the services of skilled tradesmen including a major domo who had general responsibility for organizing the work of the mill and factory, the *moledor* or miller, the sugar-master (*meastro de azucar*) who manufactured the sugar with the assistance of a staff of *espumeros*, *refinadores* and *purgadores*. Many of these tradesmen in the Canaries were Portuguese who were able to negotiate a good return for their knowledge. One major domo held a contract for 60 *doblas* of gold a year, plus 1 silver *real* for each day he spent in Castile conducting his employer's business affairs, with another 40 *doblas* depending on the outcome of the negotiations. Sugar-masters received 6% of the sugar they manufactured. The contracts for the other specialists varied, but usually specified that the owner provide food, wine and a salary of so many *maravedis* a month – 1,300 *maravedis* was common – or payment in sugar. These tradesmen had their assistants who were sometimes Africans to fetch, carry and clean, and some sugar-masters had Portuguese apprentices. The food and drink of these assistants could also be a charge on the owner of the mill (Camacho and Galdas 1961: 24–8). Not surprisingly, therefore, given these expenses, the owners of mills and factories took as their fee 50% of sugar made from the cane each grower delivered.

The relations between grower and miller were also regulated in contracts. Each, of course, was dependent on the other and each could injure the other's interests if affairs were badly managed. The miller needed to contract for sufficient cane to ensure that his mill and equipment were used to capacity but ran the risk of contracting for either too little or too much. A grower did not wish to see his cane deteriorate while waiting his turn at the mill. Growers were also concerned at the prospect of the undue profits a miller might make if the price of sugar rose substantially. Clauses therefore could be added to contracts awarding the grower a higher return than the 50% if the price of sugar sold above a certain level. The weather, accidents, a breakdown in the mill, malice or miscalculation on the part of one of the principals could upset the best arrangements so that disputes between the growers of cane and the owners of mills helped to enrich the lawyers not only in the Canaries but later in the cane-growing regions around the world (Camacho and Galdas 1961: 25–8).

The mid-years of the 1500s were the peak years of Canarian sugar production and there followed a collapse of the industry as indicated by the number of sugar mills reported on the islands (Fig. 4.2). As in Madeira, competition from the New World was the immediate basic cause, but also during these years underlying difficulties of purely local origin were becoming more acute. Fuel had always been scarce and expensive, but the progressive deforestation of the islands meant that it was gradually becoming even more so as it had to be cut at higher elevations, further from the mills. The removal of the forest cover had the further consequence of making the supply of water for the mills and irrigation more unreliable as rain water now drained rapidly away rather than being stored for more gradual release in soils and vegetation. These islands were too small, their resources too limited to support for long an exploitive form of agriculture. As in Madeira, vineyards replaced the sugar cane which soon was reduced to a garden crop of no economic significance.

São Tomé

The island of São Tomé presents an example of a sugar colony altogether different from Madeira and the Canaries, having in fact more in common with the first American colonies that were developing at about the same time, in the 1500s. In this truly tropical island, the Portuguese finally abandoned Mediterranean forms of land tenure and labor in favor of large plantations worked by African slaves. These were very significant innovations in the cultivation of sugar cane, and it is indeed unfortunate that the records of the earliest years of the colony should be so meager.

The progress of the sugar industry on São Tomé was at first slow. The appearance of a sample of the island's sugar at Antwerp in 1495 (Silva 1958: 83) did not lead immediately to continuing exports. Indeed, around 1500, a sugar industry seems scarcely to have existed, even though the Portuguese appreciated, as some contemporary comments reveal, that in the humid tropical climate they could get much better crops than they obtained in Madeira. In 1517, there were still only two *engenhos* or mills in operation. King John III gave the industry a boost when in 1529 he ordered his factor to construct twelve mills and by 1531 sugar displaced the export of timber as the island's main source of revenue. In the 1550s, there were sixty mills producing 150,000 *arrôbas* (2,203 tons) of sugar, and during the 1570s, seventy or one hundred and twenty mills, depending on which report one accepts (Silva 1958: 84–5). This increase in the number of mills would suggest a further rise in exports, but the actual figures do not survive. Estimates of sugar production based on the number of mills must inevitably contain a large margin of error, given that the capacity of mills varied greatly. There were some water mills, a few animal-powered mills, but apparently the horses could not tolerate the climate and died quickly (Malowist 1969: 15), as well as manually operated Madeiran-style *alçapremas* (Silva 1958: 88). The exports for the 1570s (Silva 1958: 84; Mauro 1960: 190) shown in Fig. 4.2 derive from their assumption of dubious validity that each ship calling at the island carried away a cargo of 20,000 *arrôbas* (294 tons) of sugar. The amounts of sugar exported are at the best estimates, but the surviving records are consistent in reporting the quality as poor. It commanded a price in Europe well below Madeiran which in itself is not surprising given São Tomé's climatic advantage but the price was even lower than that of Brazilian sugar. Once refined, however, it sold for a price rather higher than Madeiran (Silva 1958: 85–6; Mauro 1960: 190). In explanation of the poor quality, Silva (1958: 86) claims the Portuguese were experiencing difficulty in drying the sugar in the humid São Tomé climate, even going to the trouble and expense of airing the sugar over fires in purpose-built hot houses, which if indeed the case was a practice peculiar to São Tomé. But the evidence of the prices allows a different interpretation, that the planters were responding to the preferences of the refiners who found greater profit in importing the cheapest crude sugar rather than better grades for refining and resale.

In addition to the number of mills, the extent of slavery in São Tomé shows that the industry was on a different scale than in either Madeira or the Canaries. During the 1550s, when the island was well-established

60 The sugar cane industry

as a sugar colony, there were between 5,000 and 6,000 slaves, with some wealthy men owning between 150 and 300 slaves. In earlier years, ownership of fourteen slaves was considered a sign of wealth (Silva 1958: 77; Malowist 1969: 15–16). Indeed, the expansion of the industry during the first half of the sixteenth century appears to have led to a concentration in landownership and to wide disparities in wealth among planters. In its heyday as a sugar colony during the mid-1500s, São Tomé had a complex hierarchical society of wealthy and poor landowners, of priests and civil servants, of merchants from Portugal, Genoa and Flanders, of slave-traders and the slaves they brought from the Guinea coast. Soon a significant proportion of the population was of mixed racial origin. São Tomé, like so many later sugar colonies, devoted so much of its land and labor to the production of the cash crop that it could not supply its own food. Sugar was exported to Portugal and beyond; grain, wine and other supplies came back south in return. In the scale of its industry, in its dependence on slavery and in its patterns of trade, São Tomé had more in common with the later Caribbean colonies than with its immediate predecessors.

The emergence of a sugar colony that represented such a break with past experience calls for an explanation. Portuguese scholars (Garcia 1966; Silva 1958) attach great importance to the role of the humid tropical climate that the Portuguese encountered for the first time as colonizers in São Tomé. The climate did provide a more favorable environment for sugar cane. Cultivation no longer had to be confined to small irrigable areas but could extend over the countryside, "as wheat in Alentejo," to borrow a contemporary comparison, to form a monoculture. Other influences of the climate are more questionable. Some weight might legitimately be given to the argument that the hot, enervating and unhealthy conditions discouraged emigration from Portugal and so in the absence of an indigenous population, the Portuguese had to go to Africa for a slave labor force. A necessity slavery may well have been in São Tomé but less because of an enervating climate than because of a shortage of manpower in Portugal. The crown did exile undesirables from Portugal to São Tomé and even sent some 2,000 Jewish children in the hope that separated from parental control they would grow up Christian (Malowist 1969: 11), but such measures could not satisfy the demands of a growing industry. Already by 1500 the Portuguese were drawing on African slavery for labor in their own fields at home, and the crown merely extended this solution to its new colony (Boxer 1969: 31).

By 1580, São Tomé still had reserves of land and timber to allow for an even further extension of its sugar economy, but rather than growing it entered upon a precipitate decline caused by slave revolts and foreign invasions. Slaves did escape into the forested interior of the island from where they maintained a constant harassment of the plantations. There was a slave revolt as early as 1517, but those of 1580, 1595 and 1617 did much damage. In 1580, the authorities briefly lost control of the island except for the capital and its immediate surroundings. The Dutch repeatedly attacked from 1598 onwards, laying waste the plantations. Between 1641 and 1648, they occupied the island. During these difficulties, sugar production dwindled to insignificance with many of those who had been involved in the industry emigrating to Brazil (Garfield 1971: 134). Attempts by the Portuguese to revive the industry following their recovery of the island from the Dutch failed to succeed even though sugar could be produced there at a low enough price to compete with Brazil (Mauro 1960: 192). Investors in the sugar industry, it seems, required a more secure base for their capital than the turbulent lands of the Gulf of Guinea. São Tomé became a minor outpost of empire, known best to slavers who called there to collect cargoes, to repair their ships and to revictual for the Atlantic crossing.

During the mid-1500s, Europe derived its sugar from a few coastal locations around the Mediterranean, from the Sous in Morocco and from the colonies off the west coast of Africa. The rising demand for sugar in Europe meant that the market was able to absorb these new entrants without altering the status of sugar as a high-cost, luxury product. The small size of Madeira and the Canaries, the hilly terrain, the necessity for terraces and irrigation imposed severe limits on the industry's scope to expand as well as ensuring that the costs of production were unlikely to undercut those of Mediterranean producers. The frost-free climate was an advantage, but against this had to be set the disadvantage of the greater distance from markets. Nor did Madeira and the Canaries escape the problems of maintaining soil fertility, plagues of rats and insects in the cane fields and diseases of cane. In brief these islands were no threat to the European producers. The cheaper sugar from São Tomé was proof both of the advantages to the industry of humid tropical conditions and of the benefits to the refineries of importing low quality sugar. São Tomé was a harbinger of the competition to come.

The industry in America: 1492–1680

It should come as no surprise that Christopher Columbus carried sugar cane to the Americas on his second Atlantic voyage, given that he knew the industry well from his early career in the Madeiran trade, that he had married a Madeiran landowner's daughter and that he had resided for several years in the island's capital of Funchal (Morison 1942: vol. 1, 41–53; Pike 1966: 99, 154). What he saw of the Caribbean in 1492 would have alerted him to the possibilities of growing sugar cane there. The introduction of cane to Hispaniola in December of 1493 may not have been lasting

The sugar cane industry

because in those difficult early years of the colony all that was edible was eaten. Further introductions of sugar cane followed the founding of the city of Santo Domingo in 1496 (Sauer 1966: 209). Spanish explorers took sugar cane to Puerto Rico, Cuba and Jamaica in the first years of the sixteenth century, and then on to Mexico where sugar cane was growing on the coast near Vera Cruz by 1524 (Sandoval 1951: 26–7). The Portuguese also took sugar cane across the Atlantic. There is no record that Cabral had the plant on board when he reached the coast of Brazil in 1500, but we can infer that it was in the cargoes of subsequent expeditions from the order that King Manuel issued in 1516 that a sugar-master be sent to Brazil to help establish the industry there (Huetz de Lemps 1977: 50).

In introducing sugar cane to the Americas, the Spanish and Portuguese inadvertently made a valuable addition to the diet of the Indians, who rapidly discovered that sugar cane satisfied their craving for a sweetener far more satisfactorily than any of their native plants were able to do (Steward and Faron 1959: 150). They chewed the cane as a pleasant refreshment, boiled it in water to make a drink (Rouse 1949: 551) and crushed it with a simple lever-like apparatus known as "cunyaya" in which the major component was a large branch or a trunk of a tree (Marrero y Artiles 1972–84: vol. 2, 311; Morales Padron 1952: 287). The Indians' taste for sugar cane led them to pass the plant on from one group to the next well in advance of the frontiers of European settlement.

This very rapid diffusion of sugar cane was a cause of a misinterpretation of the origins and diffusion of the plant that has been very long-lived in the scholarly literature. Travelers who found sugar cane growing wild or in Indian provision grounds far from European settlements speculated about the possibility of the pre-Columbian presence of sugar cane in the Americas. Thomas Gage (1928: 19), the English Dominican, raised the issue when he came across sugar cane in Guadeloupe while on his way to Mexico in 1625. Father Labat (1742: vol. 3, 321–7), the French authority on the Caribbean sugar industry who traveled through the islands during the last years of the seventeenth century, took up the argument. He had read Gage and knew from the reports of French colonists that sugar cane was already growing on St. Christopher (St. Kitts) and Martinique as well as on Guadeloupe when the French landed to claim the islands. Labat was impressed by Jean de Léry's account (1586: 160) of the ubiquity of sugar cane in the countryside around Rio de Janeiro during the 1550s as well as by the reports compiled by Jean de Laet (1640) in his *Historire du nouveau monde* of the widespread distribution of sugar cane in the Americas so soon after the conquest. Labat was well aware of the history of the migration of sugar cane cultivation from India eastwards to the Atlantic, and accepted an eastern place of origin for the plant, but nevertheless he convinced himself that it was also native to the West Indies. The discovery during the eighteenth century that Pacific islanders cultivated sugar cane added a new dimension to the debate: if cane was not native to the Americas, it might have reached the Americas by way of the Pacific. Bryan Edwards (1793–1801: vol. 2, 204–9) reviewed the reports at the end of the eighteenth century, concluding that sugar cane was indeed already growing in the Americas when Columbus arrived with fresh samples, but he was unable to decide whether it was a native American plant or an exotic from the Pacific. Another reason for the early reports of the widespread cultivation of sugar cane may have been a simple mistake, the travelers taking for sugar cane another tall cane of the *Gramineae* family, *Gynerium sagitattum*, which was valued by the Indians of the east coast of South America for making arrows and is known in Tupi as *ubá* (Bailey 1961: 159; Cunha 1982: 303–4). The debate still surfaces in the literature from time to time (Sauer 1966: 208).

Columbus and other men with experience of sugar production in Andalusia or the Atlantic islands can have had little difficulty in recognizing the advantages for the sugar industry of their new conquests. The east coast had a humid tropical climate from Florida in the north to Florianopolis in southern Brazil while in some of the drier parts of the Spanish American mainland, as in coastal Peru, the invaders were able to usurp the Indian irrigation systems. The Indian populations rapidly died away leaving behind fertile land for the taking. They knew they would achieve much better crops than at home, and contemporary observers liked to use anatomical comparisons to emphasize just how much better they were. According to Dalechamps (1615: vol. 1, 873), around the Mediterranean, sugar cane at harvest was only the width of a human finger. Du Tertre (1654: 169) reported that Madeiran cane reached the width of two fingers, but that in the Caribbean cane grew to a width greater than the human arm. Alonso de Zuazo, one of the first sugar cane planters in Hispaniola, wrote to the Emperor Charles V of cane fields worthy of the "greatest admiration" in which cane was as thick as a human wrist and twice as tall as a man of medium stature (Marrero y Artiles 1972–84: vol. 2, 308). Another asset was the forest, as the ready availability of firewood eased one of the requirements of an industry that had heavy requirements of fuel.

The initial response of the first sugar cane planters to this abundance of resources was to abandon the conservationist practices that were hallmarks of the industry in the Old World. There was no need to spend capital on irrigation systems, no need to build terraces, no need to manure in a land where clearing new fields was less effort than striving to maintain the fertility of the old. In their absence of comment of manuring, sixteenth-century descriptions of sugar cane cultivation in America stand in marked contrast to the close attention the medieval Andalusian writers (Ibn al-Awwam 1802: vol. 1, 391; Bolens 1972b) gave the subject. The American sugar industry flourished but, surprisingly, not without a false start in the Spanish Caribbean.

Hispaniola and the early Spanish American sugar industry

The sugar industry of Hispaniola lasted for a little more than a century, from about 1515 to the 1630s. Oviedo y Valdés (1959: 106), the first historian of the industry, attributes its rather late start – about twenty years after the Spanish first introduced sugar cane – to the *conquistadores* turning "closed eyes" to the agricultural potential of the broad fertile plains of the island and giving their attention instead to enriching themselves from the gold to be found in the mines and streams of the interior. The sugar industry flourished during the mid-years of the sixteenth century when sugar accounted for up to 70% of the value of the exports of Hispaniola, but after 1580 it began to decline and within a few years had ceased to exist, a victim of circumstance and the changing course of trade and imperial rivalries in the Caribbean.

The establishment of the sugar industry came at a time of crisis in the island's affairs and was in good measure a response to the crisis. The deposits of placer gold that had been the mainstay of the economy since their discovery were, by 1515, nearly exhausted. Moreover, the island was in danger of being depopulated. The estimates of the number of Indians in Hispaniola in 1492 range from a low of 60,000 to a high of 8 million (Cook and Borah 1971: 376–410), but whatever the actual figure may have been the effects of the conquest caused such a terrible death rate that by 1520 the Indian population had declined by 90% or more and would soon become extinct. As the gold gave out, the Spaniards, too, began to abandon the island, at first for Cuba and Jamaica, and then for Mexico. The exodus continued even after the authorities threatened to confiscate the property of those who left, so that soon the only colonists to remain were the bureaucratic place-holders, in whose interest it was to stay on, impoverished landowners, old soldiers and servants, who did not have the means to pay for a passage to the Main, and the few merchants who handled the dwindling trade.

At this difficult juncture, the crown entrusted the government of Hispaniola to three Hieronymite friars for the years 1516–19 in the hope that they would be able to help the Indians. In any worldly sense this hope proved to be vain, but the friars at least came to appreciate that the island required a new economic base. There were the makings of a mixed agricultural economy in the export to nearby garrisons of maize and manioc, and to Europe of the native cotton as well as of hides from the herds of feral cattle that had now begun to roam over much of the island, escapees from Spanish settlements. The chance introduction from the east of the Cassia tree (*Cassia fistula* L.) – the Golden-Shower or Pudding-Pipe-Tree, to use its English names, descriptive of its long cylindrical seed pods – led to another export. The pods were used in medicine as a cathartic and, as *cañafistula*, appeared on the manifests of ships leaving Hispaniola from 1520 to 1545, by which time widespread propagation of the tree had led to a glut on the market. Alonso de Zuazo's experiments with the cultivation of yet another eastern import, pepper, and with the wild cinnamon of the West Indies were both unsuccessful (Cassá 1979–80: vol. 1, 170; Sauer 1966: 208–9): Hispaniola was not to be a spice island. To the Hieronymite friars as well as to Zuazo the production of sugar seemed to offer the most likely solution to the island's economic problems.

The first attempts to produce sugar had actually taken place between 1503 and 1512 near the inland service center for miners known as Concepción de la Vega. But, according to Oviedo y Valdés, the true founder for the industry was Gonzalo de Velosa who built in 1515–16 a *trapiche*, a man- or animal-powered mill, in the valley of the River Nigua a few miles to the west of the city of Santo Domingo, and at his own expense brought to his mill experts in the industry from the Canaries (Oviedo y Valdés 1959: vol. 1, 104; Sauer 1966: 209–10). Velosa thus has the distinction of being the first sugar mill owner in the Americas whose name has come down to us. The friars decided to encourage further private initiative by advancing loans of 500 gold *pesos* to those who undertook to build sugar mills. The friars realized also that the few surviving Indians could not be expected to provide an adequate labor force and so they petitioned the crown, successfully, to permit the free entry of African slaves to the island. The crown later extended a further measure of encouragement to the industry when in 1529 it declared neither mills nor slaves could be seized in repayment of debts (Cassá 1979–80: vol. 1, 84). The sugar industry was not an easy activity for colonists to enter. Would-be planters had to own or be able to borrow a large amount of capital. They had not only to acquire land, but buy slaves, carts, tools and other equipment, build a mill and boiling house and finance the entire enterprise for up to two years at least, until their first loaves of sugar reached the market. The manufacture of sugar in Hispaniola became the business of the elite. Oviedo y Valdés (1959: vol. 1, 107–10), in his discussion of the industry in the 1540s, makes the point that many of the plantations were owned by members of the upper levels of the bureaucracy or by descendants of such people. The Genoese carried their financial interest in the sugar trade to the far side of the Atlantic and in Hispaniola built and operated three of the most productive mills (Pike 1966: 129). The Hispaniolan industry also attracted funds from the Welsers, bankers to Charles V (Cassá 1979–80: vol. 1, 84–5).

Two measures of the progress of the industry are the increase both in the numbers of mills and amount of sugar exported. By 1520, three *ingenios*– large mills by the standards of the day and probably water-powered – were functioning and sugar had become one of the exports of the island. By 1527, there were eighteen *ingenios* and two *trapiches* with twelve more mills of unspecified type under construction. By the late 1530s, there were thirty-four mills, of which only four or five were *trapiches* and the number of mills appears to have remained at this level through the peak years. Sugar exports (Fig. 4.2) rose to about 60,000 *arrobas* (690 tons) a year in the 1550s to reach a maximum of nearly 100,000 *arrobas* (1,150 tons) in 1580. Production may have been larger than these official figures indicate because of contraband trade. In the 1550s sugar counted for 70% of the value of the exports of Hispaniola but, by 1580, even though a larger amount of sugar was exported, it represented only 50% of the value of exports. The coastal river valleys between Santo Domingo and Azua on the south coast were the main regions of sugar production with some nineteen of the *ingenios*. Other centers of production were San Juan de la Maguana in the interior of the island, Higuey on the east coast, Puerto Plata on the north and La Yaguana on the Gulf of Gonaïves in present-day Haiti (Cassá 1979–80: vol. 1, 85, 170; Ratekin 1954: 12–13).

The standard modern references on the Hispaniolan industry (Ratekin 1954; Sauer 1966; Cassá 1979–80) draw on such contemporary accounts as that of Oviedo y Valdés and the scatter of government and commercial reports that survive in archives, but these sources are an insufficient basis for either a detailed analysis of the finances of the industry or of its agricultural and manufacturing practices. Nevertheless, sufficient features of the industry do emerge to show that it was adapting methods of cultivation to the new environment. Landowners did not manure their cane fields nor did they irrigate them except in the dry Azua district where irrigation improved the yield, but it was cheaper than in the Atlantic islands. As was to be usual practice in the colonial American industry, hoes rather than plows were the main tools used to prepare the cane fields for planting. Sauer saw the influence of Indian and agricultural custom in the Hispaniolan habit of planting sugar cane in mounds or ridges of soil. Estates were large, and only a small portion of their area would be planted to sugar cane; the greater part was in pasture and forest with some cultivation of provision crops and other cash crops such as *cañafistula*. The owner of an *ingenio* of average capacity might have had the equivalent of 25 to 30 acres in sugar cane, employed about one hundred slaves as well as several Italian – probably Sicilian – Portuguese or Canarian foremen and technicians and produced about 4,000 *arrobas* (46 tons) of sugar annually. The wealthiest entrepreneurs in the business owned up to 500 slaves and had 10,000 *arrobas* (115 tons) of sugar to sell each year. The explanation for the rather high ratio of slaves to area of cane field, these data suggest, must lie in the fact that not all the slaves a landowner employed were engaged in producing sugar.

The Atlantic sugar industry: c. 1450–1680   67

Plate 4   An Hispaniolan sugar factory in the sixteenth century
Source: de Bry 1595: 2.

Theodor de Bry's (1595: 2) representation of work in the fields and mills of Hispaniola is perhaps the earliest illustration we have of the sugar industry in America (Plate 4). It conveys a strong visual message of a labor-intensive activity but an analysis of the scene reveals several inconsistencies that suggest de Bry was working from hearsay reports rather than personal experience and that he had a faulty knowledge of the processes he wished to depict. De Bry shows the Hispaniolan industry still relied on Old World technology. There are two mills in his picture. The one in the middle-ground is an edge-runner, a type that was common around the Mediterranean and that would have been known in sixteenth-century Hispaniola as a *trapiche*. It is powered by men, although oxen and horses were often used for this work. In the right background is an overshot water wheel that may be attached to an *ingenio* in the adjacent cabin. If so, de Bry omits to show slaves taking cane to grind in it. The second stage in extracting juice was to take the cane from mill to a press. De Bry does not show a press, although the mallet-wielding slave and his colleague holding a beam may be engaged in increasing or releasing pressure on cane in a press hidden by the mill wheel. But this is conjecture. Another possible explanation for their mysterious employment is that they are making boxes in which to ship the sugar. In his caption to the plate, de Bry claims the slaves are making sugar, but shows unorthodox methods. There is only one cauldron in which to boil the juice instead of several, and, as it is apparently continually being replenished, it is difficult to know how the juice would ever reduce to the point of crystallization. Indeed, a slave is filling the pots with contents that seem more liquid than crystallized. The filled pots are set out in a rack to allow the molasses to drain and the loaves of sugar to form – standard practice – but there is no shelter from rain. The lid-like coverings on the pots may be clay, but "claying" is more effective if done in the shade as the sun will dry out the clay. The shape of the pots, also, is unusual, making impossible the removal of the sugar loaf without breaking the pot. There are two other curious points: the cane is carried from field to mill by men rather than in carts or on the backs of beasts of burden; the cane trash is being stripped off the cane at the mill and not in the fields which would have lessened the loads to be carried. Correctly, the fuel is shown as branches or logs from trees. The scene is crowded with eighteen figures, eight of them in the mill house. The illustration is valuable if for no other reason than it is the only one from this time and place, and even if it is not an entirely faithful representation it provides an interesting point of comparison with later illustrations of the American industry.

Several factors accounted for the decline of the industry after 1580, but their relative importance is difficult to determine. The initial blow was the epidemic in the 1580s that killed a great many slaves. There was also competition from Brazil, particularly acute during the period 1580–1640 when the King of Spain was also the King of Portugal and Brazilian goods had access to Spanish markets. The changing patterns of trade and naval strategy worked against the interests of Santo Domingo and Hispaniola. The direction of prevailing winds and ocean currents meant that the fleets with the gold and silver from Mexico and Peru left the Caribbean via the Straits of Florida with the result that commercial and naval traffic declined at Santo Domingo as Havana rose to strategic prominence. The increasing irregularity in sailings at Santo Domingo discouraged the export trade. The government of Spain itself dealt another blow to the already greatly weakened industry in 1605–6 when in an effort to curtail the contraband trade in hides, it ordered the concentration of the colony's population into the eastern third of the island. Towns were abandoned, people forcibly relocated and, in the course of these events, the sugar plantations at Puerto Plata, La Yaguana and San Juan de la Maguana ceased production. Pirates and gangs of *cimmarrones* – runaway slaves – relatively secure in the unadministered parts of the island, harassed the remaining plantations. Some plantations on the south coast struggled on for a few more years, but the 100-year effort to maintain an industry was at an end (Cassá, 1979–80: vol. 1, 86, 121–4; Chaunu and Chaunu 1956–9: vol. 6, 1004–11).

The sugar industry of Hispaniola has a place in the historical geography of sugar production somewhat analogous to that of its contemporary in São Tomé: both were impermanent, but both represented a transitional stage in the evolution of the industry from the intensive small-scale farming of the Mediterranean and the Atlantic islands to the large plantations of the American industry in its colonial heyday. The long-noted gradual increase in the role of slaves in the labor force of the industry culminated in São Tomé and Hispaniola in the industry's dependence on slavery, a dependence that was to last until the abolition movements of the nineteenth century. The data from Hispaniola make clear that the scale of operations had greatly increased whether measured by the number of laborers employed, the acreage of cane fields or the amount of sugar an individual landowner annually put on the market. Production on Madeiran estates with a maximum of 22 tons was very low compared to even the production of 46 tons of an average-size estate in Hispaniola. During its peak years, Madeira produced rather more sugar than Hispaniola, approximately 1,760 as opposed to 1,150 tons annually, but from smaller landholdings. These comparisons are instructive because they lead to the conclusion that in Hispaniola the sugar industry no longer was a "democratic" business in which even small-holders could profitably participate with the labor of members of their own families but had become the "big" business of a few rich entrepreneurs who owned large properties and many slaves. Although a major crop in Hispaniola, sugar cane was never the monoculture it later became on some Caribbean islands, but shared space in the early years with food crops, Cassia trees and cattle, and in the years of its decline with cattle, ginger, cocoa and tobacco. It was the major export crop during the period approximately from 1540 to 1580 when it reached 50% and for a decade or so 70% of the value of all exports (Cassá 1979–80: vol. 1, 170). One element of continuity within these organizational changes was the persistence of Mediterranean milling technology. These various types of mills were all unsuited to dealing with the larger amounts of cane that could be grown in Hispaniola and this must have been a constraint on the growth of the industry. The largest estates of Hispaniola, those producing up to 10,000 *arrobas* (115 tons) of sugar annually, must have had two or even more mills to handle the quantity of sugar cane. The breakthrough in technology, to the three-roller mill, came too late to help the Hispaniolan industry survive.

The Spanish were even less successful in establishing sugar industries on other of their Caribbean islands than they were in Hispaniola. The few mills on Jamaica never produced enough sugar to begin an export trade (Morales Padron 1952: 287). Sugar did become the major export of Puerto Rico during the second half of the sixteenth century, but the importance of this is easily exaggerated as the island generated little trade. The annual exports of sugar varied from year to year, but seldom amounted to 10,000 *arrobas* (115 tons) (Chaunu and Chaunu 1956–9: vol. 6, 1008–9). Despite the early introduction of sugar cane to Cuba, and the ideal climate, there was little noticeable progress in establishing a sugar industry there until the late 1590s when the owners of some small mills near Havana petitioned the crown for financial aid, which in fact they received in 1602 in the form of loans. The documents from these negotiations reveal an industry that had more in common with its Canarian antecedents than with Hispaniola. There were still comparatively few slaves. The 16 landowners who took loans in 1602 had between them only 233 slaves, an average of 13.7 each, although the range was from 28 to 2. They anticipated overcoming the shortage of slave labor by a *colono* system similar to that of the Canarians and which was to become very important much later in Cuban history, whereby farmers and poor settlers grew cane which they would sell to the owners of mills. The amounts of sugar exported were small, only 5,658 *arrobas* (65 tons) from Havana in 1602, and perhaps as much as 60,000 *arrobas* (690 tons) between 1602 and 1611. By 1620, there were also a few mills elsewhere on the island, at Bayamo and Santiago de Cuba, but hides and medicinal plants, not sugar, were the foundations of the very modest economy of early colonial Cuba (Marrero y Artiles 1972–84: vol. 2, 305–21; vol. 4, 1–34).

Brazil

The Pope's division of the world in the Treaty of Tordesillas of 1493 determined that the Portuguese colonies would be in eastern Brazil which was the only part of the Americas to lie within the sphere of activity he had assigned them. After Cabral's voyage, the Portuguese returned to Brazil to cut the valuable dye woods – Brazil wood among them – and they were joined in this work by loggers from other nations. To forestall possible territorial claims by foreigners, the Portuguese changed their policy of ephemeral visits to one of permanent colonization. In Madeira and São Tomé they had learnt that sugar plantations could be successful institutions of commercial empire and they decided sugar plantations were the best means of ensuring that the new Brazilian colonies would contribute to, rather than be a burden on, the national exchequer. In their efforts to transfer the sugar industry across the Atlantic, they had an advantage over the Spaniards in that Brazil during the early colonial years offered no better outlet for investment than sugar and, what is more, no Eldorado glittered over a western horizon to tempt settlers to abandon the plantations to mine for gold. The narrow plain that fronts the Atlantic from Natal in the north to Florianopolis in the south has a climate suitable to the cultivation of sugar cane, and considerations other than climate therefore determined the specific location of the settlements. The Indians of what is now the state of Espirito Santo were hostile, able to prevent the colonization of their territory until even after the end of the colonial period. The great distance from Lisbon was a handicap to the colonists in São Vicente and Rio de Janeiro who wished to export sugar and these southern settlements remained secondary producers until the eighteenth century. The most favorable locations were in the northeast. Here, on the humid coastal plain of Pernambuco and on the fertile clay soils around the Bay of All Saints in the district that was to become known as the Recôncavo of Bahia, the Portuguese founded the most important sugar colonies of the Atlantic phase of the industry.

The sugar plantations of Brazil were called *engenhos*, a word that literally means mills, but was applied to the entire complex of cane fields, mill and factory. Landholdings were large, often measured in square leagues, and had their origins in the *sesmarias* or land grants the crown or colonial governors awarded to individuals in return for services rendered in the hope that the new landowners would encourage settlement. Typically, a plantation occupied only part of a vast estate, with much land left in forest or rough grazing, and it supported a hierarchical society from the slave at the bottom to the owner, or *senhor de engenho*, at the top. The *senhores de engenho* seldom cultivated all the sugar cane they required for their mills. The labor force, beasts of burden, carts and equipment all represented capital, and planters passed some of the burden of financing their enterprises to tenants, or *lavradores* as they were known. A *lavrador* – the word means cultivator or farmer – undertook to grow a given quantity of cane. In return, the *lavrador* received from the *senhor de engenho* a percentage, commonly 25 or 33, of the sugar made from the cane he had grown. Leases commonly ran for nine years, but some leases were for eighteen years or longer. The contracts specified the quantities of sugar cane in terms of *tarefas*, a *tarefa* being the extent of cane field that would provide sufficient cane to feed a mill for a day. It was necessarily a somewhat imprecise measure, variously defined as about twenty paces square (Piso 1648: 645–6) or thirty arm-lengths square (Antonil 1968: 156–7). Some *lavradores* contracted for as little as three *tarefas* and fulfilled their obligations with the labor of their own families; those who contracted for up to 40 *tarefas* were comparatively well-off men, with a good deal of labor and equipment at their disposal. On any one plantation, there were several *lavradores* and these tenants were responsible for growing a large part of the plantation's sugar cane. The term *lavrador* also applied to another class in Brazilian society: the cane-farmers who owned their own land. Such people cultivated sugar cane with perhaps as many as thirty or forty slaves, but preferred to sell their cane to mill owners rather than undertake the expense, risk and work of milling and manufacturing their own sugar. They received of course a higher percentage of the proceeds from the millers than did the tenant *lavradores*. There are similarities in these arrangements of plantation owners, of tenants and cane-farmers, of rich men and poor, to the structure of the sugar industry in Madeira, but the scale was on an altogether different, American, dimension.

Another indication of the size of the new scale of operations is the labor force. As on Hispaniola, the average plantation in Brazil had about 100 slaves, excluding those owned by the *lavradores*, and on the largest plantations there were double this number. From the beginnings of the Brazilian industry in the 1520s until the 1570s, these slaves were in the great majority Indians. The Portuguese did try to attract workers through barter and the offer of wages but Indian society was not structured around the production of surplus for sale or to satisfy an elite, nor did they have any tradition of accumulating consumer goods, and when the Indians had received sufficient for their immediate needs, they often stopped working. Even as late as 1583, two-thirds of the slaves on the *engenhos* of Pernambuco were Indian. Slavery was the easiest means of obtaining a reliable labor force. The first African slaves in Brazil worked as personal servants, or as technicians in the mills and boiling houses if they had had experience in São Tomé or Madeira. In some *engenhos*, they replaced Madeiran salaried sugar-masters. The increasing employment of African slaves in the fields from the 1570s was the result of a number of factors. Indian resistance to the Portuguese through warfare and rebellion continued, and although prisoners taken in a "just" war was one means of obtaining slaves, both crown and church became more firm in their opposition to the enslavement of Indians. Moreover, the ravages of introduced diseases had reduced the Indian population. The growing scarcity of Indian slaves made them more expensive than in earlier years. The Portuguese had discovered that Indian slaves were less productive than Africans, but accepted this situation as long as Indian slaves were cheaper. After the 1570s, the economics of slavery tilted in favor of the Africans, impelling a transition to dependence on African slavery that was completed during the first years of the 1600s (Schwartz 1978).

The vast plantations of Brazil presented a picture of abundant resources and profligate use that must have astonished anyone familiar with the careful husbandry of the tiny terraced fields of Madeira. The first colonists claimed the soils were extremely fertile and the forests seemingly endless. Soares de Sousa (1971: 166) writing in the 1580s stated, with perhaps some exaggeration, that around Bahia fields had been continuously cropped for thirty years while more than half a century later Piso (1648: 50) could still claim that some of the best soils of Brazil were successfully cultivated for forty or fifty years, apparently uninterrupted by fallowing. But certainly the planters made no effort to maintain soil fertility. When yields of cane became disappointing, they cleared new fields while livestock were allowed to graze old fields and cut-over land without any attempt to collect the manure and put it to systematic use. The daily cartloads of timber destined for the furnaces that Piso (1648: 51) described suggests both the scale of the demand for fuel and the shortsighted abandon with which the forests were depleted. There was ample space for the provision grounds that supplied the staples of maize, manioc and beans, all newly borrowed from the Indians, as well as yams which the Portuguese had already come across on the coast of Africa or in São Tomé. Some of the activity would have been familiar: the planting of the eyes of cane in long, parallel furrows, the weeding of the cane fields three or four times, the intense labor of the harvest and milling the cane.

During the first century of the Brazilian industry, there was no advance in the technology of the sugar mills. The *senhores de engenho* built the same types of mills of Mediterranean design with which the sugar industry had long worked, and in Mediterranean fashion supplemented the mills with presses. In Brazil, there were two or three presses for each animal- or water-powered mill (Salvador 1965: 365). The size and efficiency of a *senhor de engenho*'s mill and presses placed a limit on the number of *tarefas* of cane he could process. If he wished to increase his production he had to install more of them and find the labor to work them, that is, if he had already exhausted the strategy of lengthening the harvest season. The technology of the Mediterranean mills was ill-adapted to the scale of the industry in America and may well have slowed its rate of expansion.

The breakthrough to a more effective means of milling sugar cane was achieved during the first quarter of the 1600s in the form of a new design of mill that crushed the cane between three vertically mounted rollers or cylinders. The place and date of the invention is uncertain but the earliest-known illustration of a three-roller mill is a Portuguese sketch from Brazil (Plate 5) dated 1613 (Ajuda Palace Library).

There are two traditions about the origins of this mill in the literature. One credits the invention to a Pietro Speciale of Sicily in 1449. Von Lippmann (1929: 338) accepted this origin as did Deerr (1949–50: vol. 1, 77, vol. 2, 535) and they have been followed by several later scholars (Ashtor 1981: 106; Baxa and Bruhns 1967: 16; Barrett 1970: 53; Verlinden 1970: 20). Verlinden claimed it was a cause of the revival of the Sicilian industry in the late fifteenth century and Ashtor cited it as an example of the technological superiority of the western Mediterranean sugar industry over that of the Levant. Yet, there are very serious difficulties with the attribution

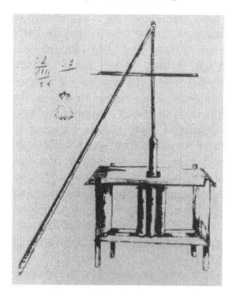

Plate 5 The three-roller mill, 1613
Source: Ajuda Palace Library.

of the invention to Speciale in 1449 (Galloway 1977: 186–7). The sources cited by von Lippmann and Deerr in support of the Speciale claim are two nineteenth-century Sicilian authors, Rosario Gregorio (1845) and Gaspar Vaccaro e Panebianco (1825–6). Pereira (1955), a Brazilian scholar, examined the writings of Gregorio and Gaspar e Panebianco as well as the sources they drew on, but he did not find any evidence of a three-roller mill. Speciale, a Sicilian official, did encourage the development of the sugar industry and built a mill or *trappeto*, but, according to Pereira, it was not of the three-roller type. Pereira concluded that von Lippmann and Deerr misinterpreted their sources, the confusion perhaps arising over the meaning of *trappeto*, a word that has been applied to different types of mills. Further doubts about the Speciale claim are raised by the fact that the three-roller mill is not reported elsewhere in the Mediterranean, the Atlantic islands or America until the 1600s. Deerr's acceptance of the Speciale claim forced him into the awkward argument that the late sixteenth-century Sicilian mill (Plate 3) was an anachronism; likewise, supporters of the claim must also explain why the Spanish on Hispaniola continued to use edge-runners. The second tradition traces the three-roller mill to Peru (Salvador 1965: 365; Mauro 1960: 204), although recent studies of the sugar industry in colonial Peru do not discuss its first appearance there (Cushner 1980; Keith 1976). The three-roller mill has similarities

The Atlantic sugar industry: c. 1450–1680 75

Plate 6 The Dutch with slaves work a three-roller mill in seventeenth-century Brazil
Source: Piso 1648: 50.

with the Chinese two-roller mill (Plate 12) and probably represents an example of the diffusion of Chinese technology to the West through the agency of the Jesuits (see ch. 8).

The three-roller mill gained rapid acceptance in Brazil: by 1618 it was already in widespread use (Anonymous 1956: 164) and ten years later had largely replaced other types of mills (Salvador 1965: 366). The speed of adoption of the mill argues that it had many advantages over the earlier equipment. The design was very versatile: it could be built in various sizes as well as be adapted to animal, water or wind power. It was sparing in its use of labor: a moderately sized three-roller mill could be worked with no more than four slaves, one on each side of the bank of rollers to pass the cane back and forth, a third to monitor the source of power and one more to bring sugar cane from the stockpile and to carry away the bagasse. Illustrations of the mill usually show the orderly activity of a few workers in contrast to the hectic scene around de Bry's Hispaniolan edge-runner (Plates 6 and 7). The new mills were so effective in extracting the juice from the sugar cane that *senhores de engenho* immediately abandoned the presses on their installation (Salvador 1965: 365). The lack of data makes it very difficult to arrive at an accurate measure of the cost-benefits the three-roller mill brought to the sugar industry but there is one calculation

76 The sugar cane industry

Plate 7 A sugar factory in the French West Indies, later seventeenth century
Source: Pomet 1694: 94.

that provides a very useful indication. Before the introduction of the three-roller mill in Brazil, each slave in the industry produced on average from 0.25 to 0.40 tons a year, and afterwards, about 0.5 tons a year (Barrett and Schwartz 1975: 542). From Brazil, knowledge of the three-roller mill spread through the agency of the Dutch to the Caribbean where the English built their mills on Barbados according to this design during the 1640s (Ligon 1657: 84). The three-roller mill soon spread to all the cane-growing regions of America and survived as the basic design of sugar mill until the nineteenth century. It was without doubt the most important innovation in milling technology that the colonial American sugar industry experienced.

A further innovation that can probably be attributed to the industry in Brazil is the adoption of a more efficient process in the manufacture of sugar. Instead of boiling the juice in the same cauldron to the point of crystallization (Plates 3 and 4), the juice was now boiled in a battery of cauldrons (Plate 7) and ladled from one to another as clarification and evaporation proceeded. There could be three, four, five, six, or even more cauldrons in a battery of which the first was the largest and each successive cauldron was smaller than the one before it. The battery was designed so that greater heat was applied to the smaller cauldrons than to the larger

(Gama 1983: 91, 157–62). With a battery, the sugar-master had greater control over the manufacture of sugar and could make the work of the boiling house a smooth, industrial activity. The battery was a necessary accompaniment to the three-roller mill and the greater scale of the industry in the Americas. The evidence of the Brazilian origin is circumstantial. Dutch illustrations of their boiling houses in Brazil do not show batteries, but Barbados, which drew on Dutch/Brazilian technology, used batteries from the beginning of its industry. Richard Ligon (1657) shows a battery of five cauldrons on the plans he reproduces of a Barbadian boiling house, and his name, "tatch," for the smallest cauldron would seem to be taken from the Portuguese for cauldron, *tacha*.

There was a slow, somewhat hesitant start to the Brazilian sugar industry in the second quarter of the sixteenth century, but thereafter it developed at an impressive rate compared to other regions of sugar production. The first exports in 1520 came from mills in Itamaracá, a colony on the northern border of Pernambuco. In 1532, there were three *engenhos* in São Vicente in southern Brazil. Exports of sugar began from Pernambuco in 1542 and from Bahia a little later after the founding of Salvador, the colonial capital, in the Recôncavo, in 1549. By 1570, there were possibly 60 *engenhos* in all Brazil, a number that increased to over 100 by the 1580s, 346 by 1629 and 528 by 1710. On the basis simply of the numbers of *engenhos*, the most important region was the coastal plain on either side of the city of Recife in Pernambuco, with about one half of the total. The Bahian Recôncavo was next in importance, but during the 1600s, sugar production in Rio de Janeiro became more and more significant until by the end of the century it had nearly as many *engenhos* as Bahia. The quantity of Brazilian exports (Fig. 4.2) soon surpassed the combined production of all the other sugar colonies around the Atlantic (Huetz de Lemps 1977: 50; Mauro 1960: 192–3).

From Brazil to the Caribbean

The role of the English, French and Dutch in the sugar industry began to change during the last two decades of the sixteenth century from that of consumers of a finished luxury product – for which they traded directly with the Spanish and Portuguese or bought in Antwerp, then the principal center of refining in northern Europe – to being refiners of sugar and ultimately owners of their own sugar colonies. The sack of Antwerp and the destruction of its refineries during the Dutch wars for independence from the Spanish Habsburgs encouraged the dispersal of sugar refining to other cities, so that by 1600 entrepreneurs in London, Hamburg, Rouen and particularly Amsterdam were well-established in this phase of the industry (Nef 1950). The Dutch developed their participation beyond refining into transporting sugar to the extent that in some years they carried about half the sugar Brazil exported to Lisbon in their own ships. In 1624–5, in a new turn of policy, the Dutch attempted to seize the city of Salvador, in Bahia, and its surrounding plantations, only to be repulsed. In 1630, they returned to attack, captured Recife which they held until 1654, and at the height of their power in Brazil managed to extend their rule over much of the coast between the São Francisco River and Maranhão. The English and French began to seize islands in the Caribbean in the 1620s and to convert them to sugar production a decade or so later.

This chronology of the growing participation of the English, French and Dutch in the sugar industry raises the question of why they seized territory in which to grow their own cane at this time rather than earlier or later. One analysis of the events provides an answer in terms of the response of these three nations to the pressures brought about by the slow-down in the world economy in the late sixteenth and seventeenth centuries. As economic opportunities became more constrained, each nation sought to prevent a decline in its share of trade through keen if not unscrupulous competition, a competition in which the trade of the weaker powers – Spain and Portugal – was very vulnerable to the encroachments of the north Europeans. In the Caribbean, for a period, this competition took the form of piracy against the Spanish. Piracy, of course, when licensed by a state and directed at the commerce of a rival state is one means of exploiting the resources of the rival's colonies without incurring the burdens of administration or defense. Contraband trade, carried on with the connivance Spanish colonists, gradually became more important than piracy as a tactic by which the north Europeans gained access to the wealth of Spanish America. A further means for the north Europeans to reinforce their own national economies was to produce for themselves the tropical goods they relied on others to cultivate. With their own colonies, they would better control the supply of goods, earn larger profits and enjoy the sanction of mercantilist theory. The imperatives were strong. Yet another rationale for acquiring land in the Caribbean is that friendly ports would help trade, illegal or otherwise. It is doubtful, indeed, whether piracy could have continued as long as it did, until the end of the 1600s, without the support of the Dutch in Curaçao, the English in Jamaica and the French in western Hispaniola. The first agricultural colonies were in the Lesser Antilles. There was a logic to this location as these islands in the early seventeenth century were uninhabited and so could be occupied without opposition from either Indians, Spanish or Portuguese. Moreover, because of their position as the most windward of the Caribbean islands, they were relatively easy to defend against attacks launched from Spanish America. They were also the closest of the Caribbean islands to Europe. When the English settled Barbados, and the French scouted out St. Kitts, Martinique and Guadeloupe, their intention was to grow tobacco and cotton. Only when they had discovered that these crops gave disappointing returns on investment did they switch to the cultivation of sugar cane. Sugar was one commodity for which the demand continued to be buoyant during these decades on the low curve of the economic cycle (Wallerstein 1980: 156–66).

The years 1630–60, when the English, French and Dutch were founding their own sugar colonies, were the peak years of Dutch influence in the industry. Their ships carried sugar from colonies to metropoles; in Amsterdam they had a major center of sugar refining, and they had markets in northern Europe. They became during these years financiers of the industry and disseminators of technical knowledge. The key to their influence was their occupation of Pernambuco which gave them not only control over the major sugar-producing region in the Atlantic world, but the opportunity to learn how to grow cane and manufacture sugar, and experience of managing plantations. The demand for sugar was such that even while their own production in Pernambuco was still increasing they started to invest in the new Caribbean colonies and to teach such potential rivals in the business as the English on Barbados how to run sugar plantations (Edel 1969). It was, as a contemporary observer wrote, "by the great encouragement of the Dutch" that Barbados became a flourishing colony (Great Britain 1880: 528). When the Dutch were finally expelled from Pernambuco in 1654, new suppliers of sugar were already in place.

In this transfer of the sugar industry into the eastern Caribbean, the history of the industry became entwined with the diaspora of the Sephardic Jews. The Dutch brought the industry to the eastern Caribbean, but the Jews followed closely behind. During the sixteenth century, Portuguese Jewry emigrated to escape the rigors of the Inquisition. Many went to the Ottoman empire where they were given a good reception, others turned north to Bordeaux and Amsterdam. Some made pro forma conversion to become what were known to the Portuguese as "New Christians" and sought refuge in Brazil where the discrete practice of Judaism did not attract the interference of the ecclesiastical authorities. In both Pernambuco and Amsterdam, the Sephardic Jews became involved in the sugar trade as financiers and merchants; in Pernambuco a few became *senhores de engenho*. Given the close bonds of religion, business and family between the Sephardic communities of Pernambuco and Amsterdam, the Jews in Pernambuco became the allies of the invading Dutch and, as a consequence, on the defeat of the Dutch in Pernambuco they had to emigrate once again. In 1654, some 600 Jews left Recife for Amsterdam. Faced with the difficulties of refugee life in Amsterdam, some preferred to return to familiar occupations, and sailed to the Caribbean where they found careers as advisers, technicians and financiers in the sugar industry. They appear also to have had a prominent part in establishing the Dutch colonies of Surinam and Essequibo along the Guiana coast. In 1659, they founded in Curaçao what was to become the Jewish community with the longest continuous history in the Americas. This community and its sisters on Barbados, Nevis, Jamaica and St. Eustatius must have been very distinctive religious enclaves in the societies of those small tropical colonies (Canabrava 1981: 36–47; Emmanuel and Emmanuel 1970: vol. 1, 7, 45; Merrill 1964; Wiznitzer 1960: 57–63).

Barbados: 1640–80

The introduction of the sugar industry to Barbados brought about a social and economic revolution that replaced a society of small-holders with one based on plantations and slavery. The English had begun to settle the island in 1627 and quickly built up a population of tobacco- and cotton-farmers who with their families and indentured servants numbered 36,000 within twenty years. Most of the families made a living from plots of less than 10 acres. Barbados covers only 106,000 acres (166 sq. m.), but in 1645, according to one account this area was divided into 11,200 properties, among them some relatively large landholdings that derived from the grants the founders of the colony made to their friends and associates (Sheridan 1974: 83, 132). This rather crowded island appeared to have poor prospects with the low prices for tobacco and cotton providing unsatisfactory incomes when, in the 1630s some of the leading citizens began to consider sugar cane as an alternative cash crop. However, conversion to the production of sugar required not only technical competence that the Barbadians did not have but also capital in quantities that few of them would be able to raise. In 1647, for instance, a half share in a 500-acre sugar plantation cost £7,000 (Ligon 1657: 22). A combination of circumstances permitted the promoters of the industry to overcome these difficulties. There was assistance from the Dutch, anxious to develop new sources of sugar supply. The Civil War in England led some Royalist gentry who had access to capital to seek asylum in Barbados, and until the Commonwealth established its authority over the island in January of 1652, the Barbadians could trade unhindered by English laws with the Dutch. Meanwhile, a scarcity of sugar in Europe caused by the warfare in Pernambuco led to a rise in price that favored those entering the industry. The Barbadians who thought they could profit from the new crop seized the opportunity the events of the 1640s offered to begin the revolution in the colony's way of life.

The concentration of landownership and the emergence of a near-monoculture of sugar cane were two basic aspects of this revolution. According to a census of December 1679, the number of landowners had declined to 2,639, excluding 405 householders in Bridgetown, the capital. The great majority of the owners were still small-holders and the median size of Barbados farm was less than 10 acres (Dunn 1969: 8), but there were also 259 holdings of 100 acres or more which together took up about 50% of the area of the island. Such estates were the dominant form of landholding in the most fertile parishes, occupying 75% of their area, but were comparatively unimportant in the northern and southwestern parishes where the drier environment was less favorable to sugar cane. During the 1640s, to generalize from Ligon's (1657) example of a plantation, a planter would have 40% of his land in sugar cane; forty years later, Sir Dalby Thomas (1690: 16) in his examination of plantation finances allocated 80 acres of a 100-acre estate to sugar cane. This figure of 80% provides a clue for estimating the extent of the monoculture of sugar cane on Barbados in the 1680s: if the plantations of 100 acres or more had 80% of their land in sugar cane, and making allowance for the cane fields on estates of less than 100 acres, then something over 50% of the surface of the island was planted to sugar cane (Galloway 1964: 40–1).

As sugar grew in significance, so did African slavery: from about 6,000 slaves in 1643 to 20,000 in 1655 and 38,782 in 1680. The contrast with the number of white indentured servants is most marked for there were only 2,317 of them in 1680, a figure that might well have been even lower were it not for the law that required planters to maintain a ratio of one servant to every ten slaves as a precaution against slave revolts and to provide manpower for the militia. Slave ownership like landownership was by 1680 heavily concentrated. The 175 planters who each owned 60 or more slaves accounted for half the slave population, but the other half was widely distributed among lesser landowners with even cultivators of no more than 10 to 20 acres owning several slaves. There was clearly in 1680 still a place for the "small man" in the Barbadian sugar economy as a cultivator of sugar cane and also as a miller. There were many owners of mills who grew only a small amount of cane and, in order to make full use of the capacity of their mill, took in the crops of neighboring cane-farmers (Dunn 1969).

A final aspect of this revolution in Barbadian society was the decline of the white population. No longer the majority by 1680, it nevertheless still stood as high as 20,000 but was to decline even further. Barbados could not retain the servants after their period of indenture was over because there was no agricultural frontier on the island to which they might go to find land of their own. Instead, they left to look for land in other Caribbean islands or in the North American colonies and some joined the crews of pirate ships. Prisoners from the Civil War in England had replenished the stock of servants, but this was an ephemeral source, although later in the century several hundred unfortunates were "barbadoed" – sent for servitude in the West Indies – for their part in Scottish troubles or in the Duke of Monmouth's rebellion (Sheppard 1977: 29).

For those who had the luck and resources to become planters, the introduction of the sugar industry was an enormous success. Profits were large, especially in the early years of high prices before exports from Pernambuco resumed their usual quota, and the big landowners of Barbados became the wealthiest men in the English possessions in America. Others – servants, small-holders and the slaves – paid the human and ethical costs. In terms of the evolution of sugar cane plantations as an agricultural system, Barbados presents the conjuncture of great dependence on slavery with the intensive cultivation of small estates, for even the largest of its plantations were small when compared to what the *senhores de engenho* in Brazil had at their disposal. This scarcity of land was in the years after 1680 to make the plantations of Barbados as well as those of the other islands in the eastern Caribbean very distinctive compared to plantations in other regions of America.

Barbados was the forerunner by a few years of a somewhat similar sequence of change on the Leeward Islands of Nevis, Antigua, Montserrat and St. Kitts – an island that the English shared with the French until they took full possession by the Treaty of Utrecht in 1713 – as well as on the French islands of Martinique and Guadeloupe although the pre-sugar phase on all these islands was far less developed than on Barbados. In 1655, the English added Jamaica to their territories and the French, who had gradually been settling the western part of Hispaniola, finally had their claims accepted by Spain at the Treaty of Ryswyck in 1697. This French portion of the island, known as St. Domingue, was to become the richest of all the Caribbean colonies in the following century. By 1680, the English and French had learnt how to cultivate cane and manufacture sugar, and they had the territorial base on which to do it. Although reliable statistics on the export of sugar from the Caribbean in this period do not exist, these islands were already becoming important suppliers of the European market. Between the years 1669 to 1700, the annual exports from the English islands rose from 11,700 tons to 27,400 tons. During the 1670s, Jamaica's contribution averaged less than 1,000 tons annually and Nevis, the chief producer in the Leewards, contributed 1,500 tons. The major share, 60% to 70%, came from Barbados.

The English and French could have taken even more territory in the Caribbean than they did. Jamaica marked the end of English territorial acquisitions in the Caribbean at Spain's expense until the Napoleonic Wars and St. Domingue was the last Caribbean colony France took from Spain. One possible explanation for the appeasement of the land-hunger of the English and French by the mid- to late-seventeenth century lies in the argument that these late entrants on the Caribbean scene now had sufficient territory from which to satisfy their demand for sugar. Moreover, the English and French sugar cane planters had a good motive for opposing the acquisition of yet more territory: new colonies would lead to increased competition in the home market and lower the price of sugar (Wallerstein 1980: 163–4).

The Atlantic period was the second important period in the historical geography of the sugar industry of the West. It was above all a period of transition from the Mediterranean world where sugar cane competed for capital, land, water and labor with many other crops, and eighteenth-century tropical America where large slave-run plantations dominated the economies of the colonies of several European powers. As the Portuguese and Spanish had colonized the Atlantic islands and then America, the industry gradually moved towards large landholdings, dependence on African slavery and monoculture, all characteristics of the eighteenth-century American sugar plantations. The agricultural techniques, too, evolved, in adaptation to different environments, most markedly in São Tomé, Hispaniola and Brazil where in the humid tropical climate and with an abundance of land, planters abandoned the conservationist techniques of Mediterranean countries. With its rapid acceptance of the three-roller mill, Brazil appears additionally as a center of technological innovation which diffused to the Caribbean through the agency of the Dutch and Sephardic Jews.

In the Americas, the climate and resources gave the industry an enormous advantage over producers in the Old World where by the end of the Atlantic period, the sugar industry had declined to insignificance. A green patch of cane on a terraced hillside, the tumbled stone-work of abandoned mills and candied fruit on the dessert table were the "madeleines" to jog the memory of Madeirans, Andalusians or Moroccans of the importance sugar cane once had in their economies. By 1680, sugar cane had become an American crop. Already some significant regional contrasts had emerged in the industry from one part of America to another and these were to become more pronounced in the succeeding 150 years.

5

The American sugar industry in the eighteenth century

During the eighteenth century, sugar cane was the most important cash crop in tropical America, dominating the economies of several colonies and making an important contribution to the economic life of many more. At the beginning of the century, the commercial cultivation of sugar cane was still comparatively restricted in its geographic distribution but by the end of the century, it was a major crop in many locations between Louisiana in the north and the environs of Rio de Janeiro and Tucumán in the south. This expansion of the industry was the result of the continuing rise in the demand for sugar, a demand that in turn was stimulated by the decline in the price of sugar. The industry remained profitable but very competitive, and planters accepted innovations intended to reduce labor costs, improve yields and mitigate the effects of deforestation and soil exhaustion. The trends towards the concentration of landownership and dependency on African slave labor, already discernible during the 1600s and even earlier, continued and strengthened. In Brazil and on mainland America, those displaced by the expansion of large estates could move to the agricultural frontier, but few Caribbean islands afforded such an opportunity and their unwanted small-holders and time-expired indentured servants joined up with either the military or the buccaneers, or emigrated to North America. The result in most cane-growing regions was a society in which a European elite of planters, merchants, military and officials dominated a small population of "poor whites" and a large population of slaves. No nostalgic gloss about leisured days and charming company in the porticoed plantation house can conceal the cruelty of the system the sugar industry created, but it was a lucrative system for those who controlled it and its legacy is still visible in the population and landscape of tropical America today.

Europe continued to be the major market for the sugar but the growing population of the Americas soon created new centers of consumption and soon there were three distinct types of markets. Europe and the North American colonies made up the international market. Transport to this

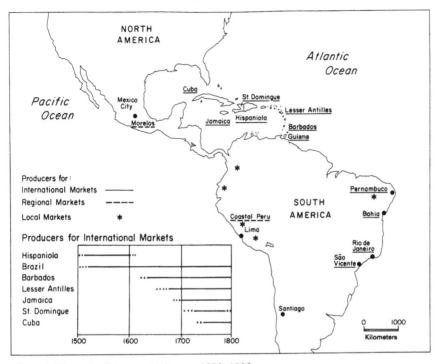

Fig. 5.1 The colonial sugar industry, 1500–1800

market was necessarily by sea with the consequence that colonies along the Atlantic seaboard from Brazil to Cuba were in the best position to supply it. Within Latin America, there were several regional markets. Mexico City and Lima were by European standards comparatively large cities and they along with other colonial cities and mining communities generated a demand for sugar large enough to support concentrations of plantations in Morelos, Mexico, as well as along the coast of Peru. The regional industries were of modest size: Morelos at the end of the eighteenth century produced about 6,200 tons annually (Barrett 1976: 163). In Brazil, sugar from the northeast as well as from a newly developed center of production along the coast immediately to the north of Rio de Janeiro found a sale in the city of Rio, in the gold-mining towns of Minas Gerais and, further afield, in Buenos Aires. In addition to the regional trade, the sugar industry also had a purely local aspect. In the interior of the continent, protected by distance from outside competition and by the same token prevented from participating in a wider network of trade, many a landowner grew cane and manufactured a small amount of sugar for consumption by the people of the immediate neighborhood (Fig. 5.1).

86 The sugar cane industry

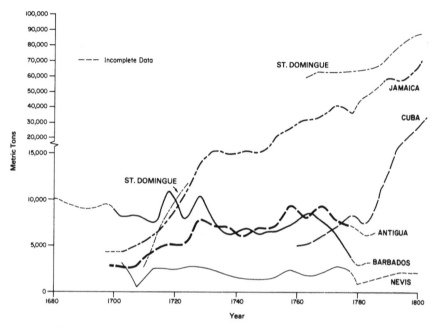

Fig. 5.2 Caribbean sugar exports, 1680–1800 (five-year averages)
Sources: Sheridan 1974: 487–92; Deerr 1949–50: vol. 1, 131, 193–9, 240.

In the eighteenth century, the Caribbean islands took on the dominant role in the supply of the international market. They had a competitive advantage over Brazil in that they were closer to Europe and North America and, moreover, the mercantilist policies of France and England effectively relegated Brazilian exports to Portuguese and Mediterranean markets. Barbados maintained its rank as the most important of the Caribbean sugar colonies until the early 1700s when it was superseded by the far larger island of Jamaica (Fig. 5.2). The other comparatively important English colonies were Antigua, Nevis and St. Kitts. After modest beginnings on Martinique and Guadeloupe, the French turned the western third of the island of Hispaniola into the flourishing colony of St. Domingue which by mid-century surpassed the output of Jamaica. The demand for sugar in Europe was able to support yet more entrants: during the second half of the century sugar began to be produced in significant amounts along the coast of Guiana and on St. Lucia, St. Vincent and Grenada, hitherto neglected islands. There had been a few modest sugar plantations on Cuba since the first days of Spanish rule but the realization of the cane-growing potential of the island did not begin until 1762 when the English briefly occupied Havana and opened it to English trade. This interlude provided a great stimulus to Cuban landowners who bought large numbers of slaves and, thus equipped, were able to expand their activities, soon making Cuba rival Jamaica and St. Domingue.

The eighteenth century marked the heyday of Caribbean colonies when they not only generated great wealth but were the hub of a commercial system involving three continents. The sugar colonies bought slaves from Africa, manufactured goods and luxuries for the planter elite from Europe and basic foods – wheat, maize, beef, pork – and timber from North America. They exported in return sugar, rum and molasses. The circumstances of the sugar colonies varied, and interesting eddies emerged in this general pattern of trade that broke the mercantilist ideal of keeping colonial commerce within the respective empires. Little of the traffic in slaves to Spanish America was in Spanish ships as a result of Spain's exclusion from Africa by the other European powers, and from the end of the sixteenth century Spain secured slaves for its possessions by granting licenses, or *asientos*, to foreign companies. The value of many an *asiento* lay not so much in the profit to be made on slaves than in the opportunity it gave for illicit trade in the ports of the Spanish Americas. Portugal, short of shipping, allowed the Dutch and English to carry much of the Brazilian trade. The French had particular problems with the trade both in food and molasses. The large, less-developed Spanish colonies were better able to supply their own food than the small British and French islands, while the British islands were easily supplied from British North America, i.e., from within the empire, but French North America could not generate an agricultural surplus of sufficient scale to perform the same service for the French Caribbean. The solution for the French islands, too, was to import from British North America. Molasses was a significant item of profit to planters, but France with its own brandy producers to protect did not wish to import this raw material for the manufacture of rum. The French planters sold their molasses instead to British North America thereby aggravating the British planters who did not like the competition. Eventually, by the Molasses Act of 1733, Britain legalized this trade, attempting to profit from it as well as to protect her own Caribbean interests by imposing a duty on the French molasses. The Act proved difficult to enforce. Indeed, the authorities of all the colonial empires found the enforcement of trade regulations in the Americas to be a problem. Economic incentives encouraged the flouting of regulations, and the large number of vessels, small and large, that plied between the islands, to the Spanish Main and along the open eastern seaboard of North America were impossible to police (Davis 1973: 283–7).

The value of the sugar colonies to the metropolitan powers is difficult to quantify. The profits of the plantations, the profits from the slave trade, the markets the plantations provided for manufactured goods all have to be taken into consideration. Some statistics will provide a measure of their importance. For England, France and Portugal sugar was by far the most important colonial import: in the case of England, each year after 1660, sugar exceeded the combined value of all other of its colonial imports; in 1774, sugar accounted for half the value of all French imports from her West Indian colonies and, over the colonial period, sugar made up a half of Brazil's exports (Davis 1973: 251). The sugar colonies not only supported a wealthy planter class but through the trade they generated provided employment for large numbers of people in Europe and North America and contributed substantially to the accumulation of capital in Europe. The basic instrument of producing this wealth was the plantation.

The eighteenth-century sugar plantation

While there was no such thing as a typical eighteenth-century American sugar plantation, planters throughout tropical America followed basically similar practices both in cultivating sugar cane and in manufacturing sugar. They faced also many similar problems. And so it is possible to provide a general description of the capital costs and work routine of the plantations.

The sugar colonies were not places of opportunity for immigrants who arrived without money or influence. A sugar plantation was a large agro-industrial enterprise that took a good deal of capital to fund and a good deal of experience to manage profitably. A would-be planter had to acquire land either through a grant if he had the right connections or in the open market if he did not. He needed a labor force which meant in nearly all parts of tropical America that he had to buy slaves. He needed also to build a mill, to build and equip boiling and curing houses in which to manufacture the sugar, and provide accommodation for his slaves, his servants and his family. He needed also carts, hoes and other agricultural implements as well as draft animals. Other sundry expenses had to be met, such as legal fees, and if the investor was founding a new plantation rather than buying a going concern he would have had to have had in addition to all these investments sufficient funds to pay for the running costs of the plantation for upwards of two years until the first returns came in from the sale of sugar from the first harvest. Ligon (1657: 116) wrote that "so small a sum as" £14,000 would set a man up in a plantation of 500 acres in Barbados, but 500 acres was by Barbadian standards a large estate and smaller plantations could be had for less. However, by 1680, the owners of large Barbadian plantations were, according to Dunn (1969: 4), "the wealthiest men in English America." In the early eighteenth century, £3,000 was enough to buy a small, run-down property on Nevis requiring additional investment to bring it into production (Pares 1950: 17), but plantations varied enormously in value depending on such factors as the extent and quality of their lands, and the numbers of slaves. In Antigua, during the mid-years of the eighteenth century, a plantation of average size by local standards, that is one of 200 acres and 100 slaves had a value of about £10,000 (Sheridan 1961: 343). During the 1770s in Jamaica, a plantation of 600 acres and 200 slaves had an estimated value of £19,027. This was a medium-sized plantation for the island, and the value of some of the largest exceeded £20,000, exclusive of the land (Sheridan 1965: 301–2).

The inventories of plantations in several parts of tropical America make the point that the slaves were by far the most important capital asset and that land and equipment made up a surprisingly small proportion of the total value. This was true of both small and large plantations. The very modest "backwoods" *ingenio* of San Juan de Buenavista, in Santiago de Cuba, had a total value in 1708 of 10,505 *pesos* of which the twelve slaves accounted for 33% and the lands for only 17.6%. The mill was valued at 629 *pesos* and the purging house at 922 *pesos*, respectively 6% and 8.8% of the total. The remaining assets consisted of equipment, livestock, a hermitage and, what might seem a bit of an extravagance in so modest a place, a residence worth 1,434 *pesos* (Marrero y Artiles 1972–84: vol. 7, 21). In the Recôncavo of Bahia during the mid-1750s, according to a document drawn up by local agriculturalists, an investor required a capital of 15,394 *milréis* to install even a "very ordinary sugar plantation." Of this sum 3,000 *milréis*, or 20%, was for land and 5,120 *milréis*, or 30%, for forty-four slaves. Accommodation was to be on a very modest scale with only 400 *milréis* allocated to house the owner, his servants and slaves. On the list of necessary equipment and livestock are items seldom found on plantation inventories elsewhere: two sailing boats to transport firewood and cane and two canoes (Pinho 1946: 138–9). The inventory of the medium-sized Jamaican plantation referred to above gave the following values: land, £6,001; slaves, £7,140; the sugar works, £3,962, with livestock and utensils accounting for the remainder of the total £19,027. This breakdown of capital costs is very similar to the one Edwards (1793–1801: vol. 2, 255) provides in his discussion of establishing a plantation at the end of the eighteenth century. His figures are in Jamaican currency: lands, £14,100; buildings, £7,000; and stock (slaves, steers and mules), £20,380 for a total of £41,480. Hall (1961: 341–2) records a rare example of an inventory, of a Grenadan plantation in 1770, in which slaves were not the most valuable asset, but this was a large property of nearly 1,000 acres that was not being fully exploited.

The rates of return on investments in the sugar industry were very uncertain. Edwards anticipated a return of 7% on the £41,480, after meeting expenses but before allowing for depreciation of buildings and equipment and the replacement of livestock and slaves. Debien (1962: 169) calculated the income on capital invested in a large plantation in St. Domingue during the last third of the eighteenth century at 5%, a rather poor rate compared to the 10% obtained on a few plantations and 8% on the return on the majority. Many of the variables that influenced the profitability of the plantations were beyond the control of their owners. War brought losses as did bad weather. Governments raised and lowered taxes and duties, and the price of sugar, molasses and rum fluctuated in a not always predictable fashion. When favorable circumstances coincided, planters could make very large profits, but when the unfavorable were in the ascendant, the planters often were forced into debt. But some control did lie in the planters' hands. The management of a sugar plantation was a very complex business that entailed exploitation of the agricultural resources of the estate and the operation of a factory. This was known to Samuel Martin (1765: 1) as "plantership" and those who did it well fared better financially than those who did it carelessly, and almost always resident owners earned a better return than the absentees who lived in Europe and employed an agent to look after their investments.

Land use

Basic to successful plantership was the careful management of the natural resources of the plantation. The practicalities of producing sugar meant that there were four principal types of land use on a plantation: fields of sugar cane – cane-pieces as they were known to the English – provision grounds to provide food for slaves, pasture for the livestock and a forest reserve to supply building timber and to ensure fuel for the boiling houses. Because of the costs of transporting sugar cane and fuel, the ideal organization of land use was in concentric circles around the mill and boiling house with the cane-pieces at the center, then forest, provision grounds and finally pasture. On a plantation where bagasse had replaced firewood as the fuel, the ring of forest could be relegated to the periphery or even eliminated altogether. Topography, differences in soil fertility and other factors such as land tenure meant that the ideal can rarely, if ever, have existed, but the basic principle of organization held true: the cane-pieces occupied the most accessible parts of the plantation and the mills stood close or amongst them to minimize transportation costs.

Successful plantership required careful planning of the planting of the cane-pieces. The two key constraints in drawing up any plan of planting were, in addition to the availability of land, the capacity of the mill and the supply of labor, for to grow more cane than the labor could harvest or the mill crush was both a waste of effort and an uneconomic use of land, while to grow too little cane meant that expensive investments were not being optimally used. Managerial skill lay in staggering the planting of the cane-pieces so that all the sugar cane did not mature at the same time and so overload the mill and the work force. A harvest season that lasted several months enabled a planter to process a given quantity of sugar cane with a smaller labor force and smaller mill – in other words with a smaller capital investment – than a season limited to a few weeks. Ratooning complicated the planning. The first crop, or plant cane, matured in fifteen to seventeen months, but the ratoon crops were ready for harvest after about twelve months. Moreover, a planter might be wise to grow only one or two ratoon crops from each planting or several depending on yields he was obtaining and the labor he had at his disposal for planting cane-pieces. By the beginning of the eighteenth century, planters had taken to laying out the cane-pieces in large squares of equal size in so far as the topography permitted (Plate 8) in an arrangement that enabled them to organize more carefully the sequence of plant and ratoon crops as well as to monitor the yields more accurately. There were other advantages. Overseers could check easily on the work of the slaves. The paths separating the cane-pieces served as fire-breaks; they also admitted carts into the midst of the crop without having actually to drag them onto the cultivated land where the wheels and feet of the oxen would damage the roots of the ratoons. The paths could, if necessary, be planted with food crops during the early stages of the cane-cycle. To Samuel Martin of Antigua (1765: 34) the aesthetics of the regular layout was also one of its attractions. However, once decisions were implemented, planters did well to stick to them because even a temporary disruption in a programed pattern of planting and harvesting could adversely affect sugar production for a number of years.

The cane-pieces required a good deal more attention than the other forms of plantation land use and here again the quality of plantership had an influence on yields and profits. Well-managed field labor hoed and planted more carefully, weeded more carefully than the poorly managed. Manuring could raise yields but added to costs, and hence the decision when to manure was not one to be taken lightly. Pest control, too, called for skill and judgment. Rats, for instance, were a common pest, gnawing on and so damaging the cane, and an infestation could be a serious problem. Some planters employed elderly slaves as rat catchers but a more effective antidote was to set fire all around the periphery of the cane-piece and allow it to burn inwards, thus trapping and killing the rats. But fire was both a menace as well as a useful tool. A "hot" burn as like as not set by a slave in a show of protest, out of control, fanned by winds, in cane not yet ready for harvesting, could cause great financial loss; but a controlled, pre-harvest burn that cleared away leaves, snakes and rats made the work of harvesting lighter without reducing the moisture content of the cane (Barnes 1974: 344–8). There were matters, of course, over which

92 *The sugar cane industry*

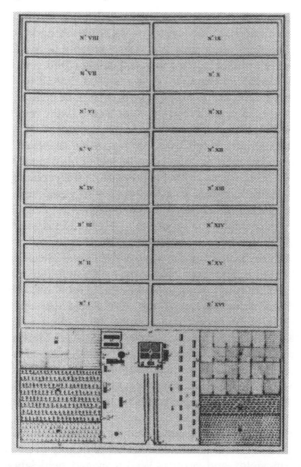

Plate 8 The ideal layout of a sugar plantation in the French West Indies: sixteen cane-pieces leave little space for the mill, boiling house, great house, slave quarters and provision grounds
Source: Avalle 1799.

plantership had little influence. Like all farmers, planters complained of the weather; certainly there were good years and bad, and the rare hurricane did great damage to the buildings as well as to the crops standing in the fields. Disease in the cane was a risk, but the frequency and significance of the occurrences is now difficult to assess. However, given there was no alternative to Creole cane until the late eighteenth century, planters accepted their losses, and if necessary cleared the land to cultivate anew.

A planter seeking to maximize his income from sugar allocated as much land as possible to cane consonant with maintaining a minimally adequate supply of provision crops, animal feed and fuel. Where land was abundant and planters owned extensive estates, there was no competition between these various forms of land use. Moreover, the planters had no particular difficulty in maintaining the fertility of the soil as they could fallow the cane-pieces or even practice a long-term rotation between cane-piece and forest. As long as the planters had access to an abundance of natural resources they were not under any imperative to modify their agricultural strategy. By contrast, in regions where resources were scarce and plantations small, planters faced many complex managerial decisions. Their aim, of course, was to produce sugar, but they could only increase the extent of their cane-pieces by taking land from other uses which in turn brought difficulties in the supply of food, fuel and pasture. This competition between the various land uses was complicated by the need to counter the effects of the depletion of resources. Deforestation cleared land for crops, but it reduced the supply of fuel. Planters without sufficient land to permit fallowing had to apply manure to maintain the fertility of the soil and so the demand for manure became a major factor in determining the numbers of livestock and the area of pasture. Given these constraints in the management of plantations, planters had continually to reevaluate their allocation of land to achieve the optimum – that is the most profitable and ecologically sound – combination of uses. Understandably, planters were open to innovations that would help them manage their estates more effectively.

The literature records a variety of innovations, some more important than others, some better documented than others. They include changes in agricultural practices and in the use of natural resources, the adoption of exotic food plants for the work force of the plantations, a new variety of sugar cane, a new design of furnace and modifications to the three-roller mill. Planters adopted innovations more readily in some parts of tropical America than in others, and the willingness of planters in a given colony to innovate varied over the years. Brazil was a center of innovation during the early decades of the American sugar industry but thereafter Brazilian planters lapsed into a traditionalism that by the eighteenth century had become the target of the criticism of agricultural improvers (Galloway 1979). Barbadian planters, on the other hand, first borrowed techniques from Brazil, then began to make their own innovations during the years around 1700 and continued to adopt innovations during the eighteenth century. With the passage of time, the contrasts between the land use not only of Brazilian and Barbadian planters but also between innovative and less innovative or traditional planters elsewhere in tropical America became more marked and the industry took on distinct regional characteristics. With the assistance of a verbal model, Galloway (1985) provided an explanation of the diffusion of these innovations that accounts both for their adoption in time and place as well as for the persistence in some colonies of traditional practices.

The adoption of innovations

The model accepts that American sugar planters as a class had two fundamental, easily substantiated, characteristics: (1) that planters had access to information about innovations and (2) that planters only adopted innovations when it made good economic sense to do so. Individual planters did of course make bad decisions and so lost money if not their plantations, but the concern here is that the planter class as a whole was informed and made rational business decisions.

No doubt much knowledge of innovations spread by demonstration and word-of-mouth, and of this there is little documentary record except of the discussions that took place in the economic societies of several Spanish American cities (Shafer 1958). But there is abundant evidence in the contemporary literature (Ligon 1657; Sloane 1707; Chélus 1719; Leslie 1740; Labat 1742; Beckford 1790; Edwards 1793–1801) of debates about the merits of particular innovations. Samuel Martin's *Essay on plantership* passed through seven editions between 1750 and 1785, and the seventh edition was reprinted in 1792 and 1802 (Sheridan 1960: 138). The metropolitan governments promoted the cause of agricultural improvement in their colonies. Even the conservative Portuguese reassessed the agricultural potential of Brazil, arranging for the distribution of seeds and "know-how" pamphlets (Galloway 1979), while at the other end of the political spectrum, the revolutionary French reviewed the state of their sugar industry (La Couture 1790; Avalle 1799). All the metropolitan governments sponsored scientific expeditions in which the search for plants of economic value was an important motive, and exotics to stock new American botanical gardens often arrived on board naval vessels. The voyage of H.M.S. *Bounty* to Tahiti to collect breadfruit for planting in the West Indies attests to the interest of the British government in finding a cheap food for slaves. These diverse activities strongly suggest that neither lack of information nor lack of official encouragement were obstacles to innovation in the sugar industry.

The characterization of the planter class as having been rational in its economic behavior should be easy to accept. A colonial plantation did, after all, represent a large investment of capital, and the defense of investment was normal behavior of mercantile capitalists in early modern Europe. The success of the planter class in defending its interests is confirmed by the fact that of the various sugar colonies only two went out of business before 1800 and they failed for reasons other than the bad judgment of planters. Metropolitan politics, warfare and other factors can reasonably be said to have brought about the end of the Hispaniolan industry, while in St. Domingue the industry was destroyed at the end of the eighteenth century by a successful slave revolt that was inspired in part at least by the French Revolution. Acceptance that the planter class acted rationally is important to the argument because it means the adoption of innovations was not random but the result of consciously taken decisions. Innovations in the sugar industry cost money: new agricultural techniques could add to the expense of production by increasing the demand for labor. This demand could rise steeply, as in Jamaica, when it became necessary to manure the cane fields. There according to Leslie (1740: 337) "100 Acres of Cane require almost Double the Number of Hands they did formerly, while the land retain'd its natural Vigour." The adoption of exotic plants involved costs as planters had to acquire seed stock and, except in the case of chance introductions, there were at least some hidden costs to the industry as planters paid through taxes and duties towards the expenses of the naval and scientific expeditions that brought the plants to America. However, a planter class that made rational business decisions can be expected to have adopted innovations only when there was a good expectation of realizing a return on the investment.

In addition to the economic rationality of the planter class, the two other components of the model are the availability of natural resources and market forces. The depletion of resources, such as a decline in the fertility of the soil or deforestation, forced planters to adopt innovations to save themselves from the deteriorating situation; by contrast, the abundance of resources permitted them to continue to cultivate their plantations without the necessity of making changes in their routine. The role of markets as a variable influencing the adoption of innovations became more important as different types of markets emerged. At first there had been only one market for American sugar: the international one. This situation changed as the colonial population grew, developing local and regional markets within the Americas (Fig. 5.1). Local markets discouraged the adoption of innovations because they were small, self-contained and usually well inland, and the transportation costs that protected producers in these markets from outside competition also prevented them from exporting to larger, more competitive markets. There was little point in producers who served these markets investing in improvements. Regional markets, represented by large cities such as Mexico City and Lima, afforded planters better possibilities for a return on investments and so encouraged an element of competition and agricultural improvement. The largest market of all was the international one. Competition was most intense in colonies catering to this market, the search for means of cutting costs and improving yields the most persistent.

The three hypotheses in the model, tested and sustained (Galloway 1985), can now be accepted as statements describing the adoption of innovations

1. Innovations that conferred high net benefits on all planters independently of resource and/or market considerations rapidly diffused throughout the industry.
2. The scarcity of natural resources as well as the depletion of resources encouraged the adoption of innovations.
3. Competition in the market place encouraged the adoption of innovations.

The first illustrations of statement No. 1 are to be found in the early years of the industry in the Americas when the planters abandoned the Old World conservationist techniques of irrigation, terracing and manuring that were unnecessary amidst the New World abundance of land. The three-roller mill and the battery of cauldrons in the boiling house are the two outstanding instances from the seventeenth century. The best eighteenth-century example of this type of innovation is the introduction to the Americas of Otaheiti or Bourbon cane – both names are used. It had three important advantages over the Creole variety, which had been the only variety of cane grown in America since the beginning of colonization. First, it gave better yields of sugar per unit area of cane-piece than did the Creole. Humboldt placed the improvement as high as one third in Mexico and one quarter in Cuba (1811: vol. 3, 15; 1826: vol. 1, 235). Secondly, Otaheiti bagasse made a better fuel than Creole bagasse (Humboldt 1826: vol. 1, 235; Fraginals 1976: 86). Thirdly, it matured more quickly than the Creole (Fraginals 1976: 86). With these advantages, its rapid acceptance by American planters is hardly surprising. Bougainville, the French navigator, found the cane in Tahiti and took samples to Mauritius, then known as the Ile de Bourbon, in 1768. It reached the French West Indies about 1780. By 1793, some samples were growing in St. Vincent, presumably in the botanical garden the British had founded thirty years before. In 1795, a Jamaican botanist obtained a sample for his island from Santo Domingo, but in the next year Captain Bligh brought more samples directly to Jamaica from the Pacific. The plants brought by the French and by Captain Bligh provided the parent stock from which the Otaheiti cane in America descended (Barnes 1974: 4; Brockway 1979: 74–5). Otaheiti cane had reached Mexico, where Humboldt saw it growing, by about 1800 and French Guiana by 1811, if not earlier, because in that year samples were sent from Guiana to Brazil (Galloway 1979: 776–7). There was initially resistance to the adoption of Otaheiti cane on the part of some planters with small or poorly constructed mills who found the thicker Otaheiti cane difficult to process (Fraginals 1976: 86), but by the 1820s it was well-established, and shortly became, if it was not already, the principal cane of the American sugar industry (Fig. 5.3).

The innovations that reflect the roles of the depletion of resources and market forces on the land use strategies of the planter class can be divided

The American sugar industry in the eighteenth century 97

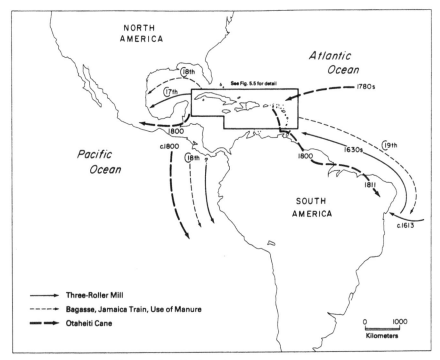

Fig. 5.3 Diffusion of innovations in America

into two categories: those that gradually diffused as conditions warranted through the greater part of tropical America, and those that were adopted in response to the particular conditions of one or two colonies.

Scarcity of fuel had been a persistent problem of the sugar industry except during the early years in America but as the industry expanded and the forest gave way to fields, in many colonies firewood once again became an increasingly scarce commodity. The solution to the problem lay in the substitution of firewood by bagasse, or mill trash as it was also known in the English islands. It was taken from the mill, dried in the sun and stored for future use in shelters or "trash-houses" that became standard features of the plantation landscape as the forest receded. The introduction of the three-roller mill was critical to this substitution because rather than pulping the cane stalks it left them crushed but intact and combustible when dried. A second factor was the introduction of a new design of furnace known as the Jamaica train. It improved on the existing design that had a separate fire under each of the several cauldrons of the battery by conducting the heat of one fire to the cauldrons through internal flues, achieving thereby efficiencies in fuel consumption (Fig. 5.4). These two innovations

98   *The sugar cane industry*

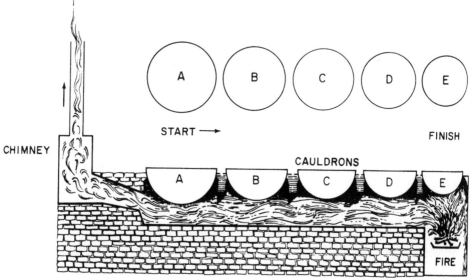

Fig. 5.4 The Jamaica train
Source: after Gama 1983: 162.

helped planters towards self-sufficiency in fuel, and both remained standard equipment until well into the nineteenth century.

Contemporary observers provide the evidence that the planters of Barbados were the first to turn to bagasse for fuel, sometime after 1650. Richard Ligon (1657), who witnessed the establishment of the sugar industry on Barbados during the 1640s, makes no mention of the use of bagasse but Sir Hans Sloane (1707: 34), visiting the island forty years later, noted that fuel was a problem, that planters "wanted wood" and had begun to burn bagasse. Labat's observations in the 1690s indicate the progress of the substitution through the Lesser Antilles. By the late 1600s, bagasse was already the fuel of the boiling houses in Barbados, Nevis and St. Kitts as well as on the "basse-terres" of Martinique and Guadeloupe. On the higher, inland locations of these two islands, planters were still able to draw on the remnants of the forest for firewood (Labat 1742: 405–6). The adoption of bagasse as the fuel may have been complete throughout the Lesser Antilles by 1719 since Chélus (1719: 159) referred only to bagasse when he discussed the use of fuel in the sugar industry there. On the Greater Antilles, shortage of firewood did not become a problem until the mid-years of the eighteenth century. The forest was of course more abundant on these larger islands and production of significant quantities of sugar began comparatively late, after the 1660s in Jamaica, after 1700 in St. Domingue and not until after 1750 in Cuba. The expansion of the industry in all three colonies led to an eventual shortage of firewood. By the 1750s, the planters in St. Domingue had turned to bagasse (Debien 1962: 42) but in Cuba sufficient forest remained to provide firewood until the end of the eighteenth century (Arango y Parreño 1952: vol. 1, 127). In northeast Brazil, the switch to bagasse came even later, after 1800 (Galloway 1979: 767–8).

The acceptance of the Jamaica train was almost contemporaneous with the changeover in fuel. In the late 1600s, the furnace was still a novelty among the English planters, but by 1725 the French were using it too (La Couture 1790: 129). The Cubans followed after about 1780, the Brazilians after 1800.

The record therefore shows that these two linked innovations, the use of bagasse as fuel and the Jamaica train, appeared first in small islands, of limited resources, and which were competing in the international market, then were adopted in the larger Caribbean islands as they too were drawn into the international market and, finally, were accepted in Brazil where forest was abundant only after the end of the colonial period. The diffusion of another innovation – the use of manure – reinforces this pattern.

The simplest way to deal with the problems of declining yields of sugar cane caused by the loss in the fertility of the soil was to practice a form of shifting cultivation, abandoning old fields to rest and recuperation, clearing new land as needed from the forest. This was the strategy the first sugar planters in any colony followed, but one that was practical only for a brief period, if at all, on the small Caribbean islands. Complaints about declining fertility in Barbados began as early as the 1650s and soon thereafter manuring became a standard part of plantation routine. Jamaican planters did have more land at their disposal than the Barbadians, and in the early years of Jamaica's sugar industry a planter might crop only 500 out of his 3,000 or 4,000 acres (Leslie 1740: 25). However, by the 1680s they had begun to prepare for the inevitable, not yet manuring their land but apparently storing dung for the future, "seeing," as Sir Hans Sloane (1707: xlv) put the matter, "by the example of their Neighbours in Barbados, that they might need it." The English writers of the eighteenth century emphasized the necessity of manure (Leslie 1740: 334–5; Martin 1765: 7, 15–16, 19–25; Beckford 1790: 204; Edwards 1793–1801: vol. 2, 220–4), discussing in some detail the virtues of different types of manure and recommending treatments appropriate to particular soils, one even doing so in verse:

> PLANTER, if thou with wonder would survey
> Redundant harvests, load thy willing soil;
> Let sun and rain mature thy deep-hoed land,
> And old fat dung co-operate with these.
> (Grainger 1764: 19, lines 197–200)

The sugar cane industry

Among the improvements the writers recommended were the addition of marl or sea sand to clay soils, the addition of clay to sandy soils, as well as the application of seaweed, cane trash and various types of dung. Martin (1765: 15), for instance, particularly favored spreading sheep dung on the cane-pieces. As manuring became a necessity, so livestock became more prominent in the plantation economy and compost heaps and animal pens made their appearance in the landscape. In contrast to this intense activity in field preparation by the English, Cubans were able to continue until late in the eighteenth century the long-term rotation of field with forest (Arango y Parreño 1952: vol. 1, 125–6; Fraginals 1976: 20). Throughout the colonial period, the planters of Morelos, who supplied the regional market of Mexico City, did not dung their land (Barrett 1970: 45, 69; Barrett 1979: 18–19), nor does Cushner (1980) mention manuring in his discussion of the cultivation of sugar cane in Peru. In northeast Brazil, also, the extensive approach to land use continued well beyond 1800 (Koster 1816: 336–63).

In several regions of tropical America, the physical environment was either too wet or too dry or in some other way fell short of the ideal for the sugar industry. The effort the planter class made to overcome such difficulties was most persistent where there was both a shortage of land and the possibility of recouping the investment that the appropriate innovations entailed. The good location of the West Indies with the ease of access to the European markets for sugar justified investment in measures to overcome their disadvantages of small size and difficult environment. The above discussion shows Barbados as having been an important center of innovation. Two more innovations reinforce the importance of this colony's place in the historical geography of the sugar industry (Fig. 5.5).

Sheet erosion and gullying became serious problems in Barbados after the deforestation of the 1600s. In response, the Barbadians adopted the practice of cane-holing (Watts 1968: 144). Instead of planting the cane in trenches, the slaves marked out the fields in carefully aligned 5-foot squares. Within each square they dug a depression approximately 2 feet by 3 and 5 to 6 inches deep, raising the soil along the tops and sides of the depression into ridges known respectively as saddles and backs. The ensemble constituted the cane-hole. The cane was planted in the depressions, packed about with manure. The checkerboard of cane-holes prevented the rapid run-off of rain water, thereby checking erosion and preserving moisture. It is still uncertain how widely this innovation diffused and for this, the loose and imprecise use of agricultural terminology by the eighteenth-century authors is to blame. For many writers, to hole land became synonymous with preparing it for planting sugar cane even if the cane was to be planted in trenches. Bryan Edwards wrote of holing land with plows (Edwards 1793–1801: vol. 2, 214). Out of this rather confusing

The American sugar industry in the eighteenth century 101

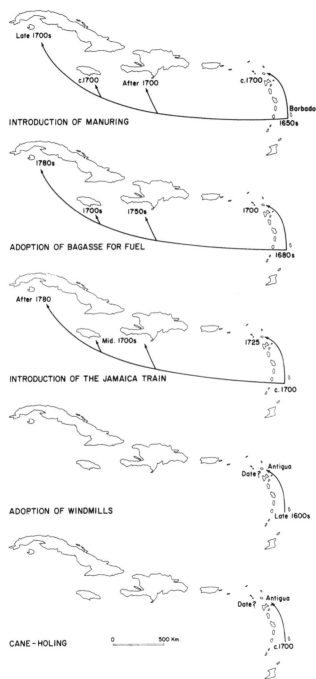

Fig. 5.5 Barbados as a center of innovation

situation in which references to holing and to cane-holes do not necessarily imply the existence of the distinctive Barbadian cane-hole, there emerges good evidence only for its adoption by the planters of Antigua (Martin 1765: 32; Clark 1823: plates 2 and 3). This limited diffusion suggests that the control of erosion and conservation of moisture may not have been the entire rationale for the technique. Barbados and Antigua are both small, mostly coral limestone islands in which conservation of soil and moisture were imperative. But the limited area of the islands also meant that land use had to be intensive. The saddles and backs of the cane-holes gave access to the fields for weeding, but they could also in the early stages of cane-growth carry a catch food crop. Cane-holing, therefore, had the advantage of permitting food crops to be grown on the cane land for at least a part of the cash crop cycle. Cane-holing, Barbadian style, was tedious, labor-demanding work to which planters resorted only when environmental constraints combined with market opportunities made it rational to do so. The practice continued in Barbados until the 1960s when it was abandoned because of the dwindling supply of labor (Barnes 1974: 281; Hudson 1984: 14).

The physical geography of Barbados did confer one advantage to the planters: its low-lying terrain in the path of the trade winds permitted them to use windmills, the transition from animal-powered mills taking place during the late 1600s (Sloane 1707: 33). The source of power for the mills had to be reliable otherwise planters risked the spoilage of their crops, and only on Antigua, another mainly low-lying island in the path of the trade winds, were planters able to follow the Barbadian example. Elsewhere in tropical America windmills were rare, even on other islands of the Lesser Antilles: there were, for instance, only eight windmills on Martinique in 1732, five in 1762, and of the 252 mills on Guadeloupe in 1739, only two were windmills (Bibliothèque Nationale: a). Windmills did not free planters of the problems of maintaining livestock because they still required manure and beasts of burden, but they did give the planters a little more room in which to juggle the production of fodder with the other demands of labor and land use.

In St. Domingue, the relatively low rainfall in the only two extensive areas of level land – the Artibonite Plain and a plain known as the Cul-de-Sac – restricted the sugar-producing potential of the colony. The access the colony enjoyed to the French market justified attempts to increase sugar production by investing in irrigation systems. Plans for irrigating the Artibonite Plain were first proposed during the 1740s, but they were never realized. To judge from Saint-Méry's discussion of the subject, the problems lay in a combination of wavering government support, in engineering and financial difficulties as well as in disputes between landowners over water rights (Saint-Méry 1797–8: vol. 2, 128–55). An amateur cartographer caught the disappointment this failure caused when he filled in the blank of the Artibonite Plain on his 1778 sketch map with the laconic inscription that this plain should be irrigated (Bibliothèque Nationale: b). In comparison, the irrigation of the Cul-de-Sac was a success. Work began in 1730 to proceed in a fitful fashion until the 1780s by which time rather more than one third of the area of the plain was irrigated and the number of plantations had increased from 3 in 1713 to 118 (Saint-Méry 1797–8: vol. 2, 267–82). There was success also in irrigating the smaller Plaine des Cayes on the south coast after 1750 (Saint-Méry 1797–8: vol. 1, 668–85), but elsewhere in the colony irrigation appears to have been very limited, the achievement of individual planters or of a few neighbors exploiting small streams (Saint-Méry 1797–8: vol. 2, 117; Debien 1962: 33, 38–9). The French were the only sugar planters in the Caribbean to make significant use of irrigation, exploiting possibilities Edwards (1793–1801: vol. 3, 166) ruefully noted that nature had "freely bestowed on many parts of St. Domingo" but had "denied to the lands of Jamaica, except in very few places."

The use of irrigation in Morelos, Mexico, and in coastal Peru might appear to be a significant discrepancy from the general explanation of the adoption of innovations: irrigation, a means of intensifying land use, is unexpected in sugar industries that catered, as these did, to regional markets. There are two points to make. Both Morelos and coastal Peru have long dry seasons and irrigation was essential if sugar cane was to be cultivated, even if only to supply regional markets. Secondly, sugar cane occupied only a small portion – as low as 5% – of the area of the estates. Much land was in pasture and other crops (Barrett 1970: 27, 33; Cushner 1980: 63–5). Therefore, despite the presence of irrigation, owners were not in fact using their resources intensively to produce sugar cane. Locked into a regional market, there was no justification for planters to do so.

The Guiana coast provides another example of the advantages of location with easy access to the international market justifying investment in innovations to overcome environmental problems. The Dutch drained and reclaimed land through systems of dykes and canals to convert part of this coast into prosperous sugar colonies. During the second half of the eighteenth century, a distinctive cultural landscape emerged in which plantations reached in long narrow strips from backland to sea wall or river bank so that owners could control the flow of water across their properties and move cane from field to mill by boat rather than by ox-cart (Robinson 1983: 245).

The problem of feeding slaves and the plantation livestock gave rise to innovations in the form of plant introductions. Some of the economically useful plants arrived in the Americas by chance, others came as part of the provisions of ships and some were deliberate introductions. During the early years of settlement, the indigenous American food crops were supplemented by the arrival from the Orient and Africa of bananas, various yams, eddoes and other *Colocasia*, often through the agency of the slave trade. Some introductions occurred surprisingly late: akee, now considered a traditional Jamaican food, came from West Africa in 1778 and mangoes only reached Martinique in 1782, on board a French naval vessel (Parry 1955: 16–17). The chance introduction of Guinea grass from West Africa was a boon to planters in those colonies where there was pressure on resources because it is a hardy nutritious plant that cattle can graze on or it can be cut for hay and pen-feeding. In Jamaica, where it was first reported during the 1740s, it soon became an important fodder crop (Moreton 1793: 58–9; Ormrod 1979: 166), but it did not even reach Brazil until 1820 when an "improving" Bahian landowner managed to acquire seeds (Pontes 1976: 179–80). Through the introduction of Otaheiti cane, the sugar industry benefited in a major way from the European exploration of the Pacific; the English tried to make it benefit also from a Polynesian staple food: breadfruit. The mutiny on the Bounty ended the first attempt to bring samples from Tahiti to the Caribbean, but a second expedition succeeded. However, the slaves did not take kindly to this food, and they and their descendants did not overcome their distaste for it until well into the nineteenth century (Parry 1955: 19). In the long run, rice proved the best answer to the problem of cheap food for plantation laborers. A seventeenth-century introduction to the Americas, it became an important crop during the 1700s in the Carolinas (Clifton 1981: 266–83) and in Brazil where the Portuguese government actively promoted it. It has since continued to gain in importance.

This international diffusion of plants did make for a more varied diet for those who had the land on which to cultivate them. Planters in Cuba, in Brazil, Mexico and elsewhere on the American mainland did have the land, but on the smaller Caribbean islands, where the economic pressures of international competition forced planters to concentrate on keeping as much of their land in sugar cane as was possible, the introductions were no solution to the problem of feeding slaves. These islands slipped into what was to become the classic colonial syndrome of exporting a cash crop and importing food.

The model shows that there was a logic to the diffusion of innovation in the colonial American sugar industry. It provides an expected pattern for the adoption of innovations in all the sugar colonies: planters adopted rapidly the innovations that conferred high net benefits independently of resources and markets and adopted selectively the other innovations in accordance with the availability of resources and market forces. If changes occurred in the resource base or in the nature of the market, the model adjusts accordingly the expected pattern of innovation. It makes comprehensible the very different records of innovation in, for instance, Brazil where an abundance of resources and a gradual retreat from intense competition permitted traditional methods to persist into the nineteenth century, in Barbados where scarce resources and competition in the international market brought early and rapid adoption of innovations, and in Cuba where innovations came in the latter half of the eighteenth century as the island began to produce for the international market. By the end of the eighteenth century, the American sugar industry was already 300 years old and, as in all agricultural systems, its land use had evolved – more rapidly and far-reachingly in some colonies than in others.

The mills

In contrast to the great level of change in agriculture and in the management of the natural resources of the industry, milling technology was extremely conservative, with only minor modifications being made to the three-roller mill. One labor-saving device was the "trash-returner", also known as the "dumb-returner" and the *doubleuse*. It was a curved metal plate that was installed along one side of the rollers to guide the cane back between them for a second crushing, and it freed one or perhaps two slaves for other duties. The hope of extracting more juice from the cane prompted some experimentation and two techniques were thought worthwhile, at least by some planters. One was to strengthen the rollers with hoops of iron, the other was to increase the diameter of the lateral rollers so that they rotated at a different speed from the central one, thus increasing the friction on the cane. It is difficult to know how successful these techniques were or how widely they were adopted. Edwards (1793–1801: vol. 2, 228–9), for one, had scant respect for them, considering that the efficiency of a particular mill in extracting the cane juice depended less on "these supposed improvements" than on the skill of the millwright who built it.

The manufacture of sugar

The work in the boiling house had to be coordinated with that in the mill and, one move further back, with the harvesting to avoid either the waste of cane juice or time-consuming delays. From the mill the juice was carried, or, if the terrain permitted, flowed along a pipe, to the boiling house where it was held in a large container called the receiver while awaiting the attention of the workers. In a well-run plantation, the juice remained in the receiver for few hours at the most, and some authorities would allow no more than an hour, because the juice quickly starts to ferment and fermented juice will not crystallize. There was indeed no point in milling cane more quickly than the boiling house could cope with the juice and it was not unusual, according to Labat (1742: vol. 4, 57), to stop the mill to allow the boiling house to catch up. Alternatively, for a large boiling house to operate efficiently, the product of two mills could be required (Barrett 1965: 157).

The basic feature of the boiling house was the row of copper cauldrons, known as a battery, set over a furnace or furnaces. It was normal for a boiling house to have only one battery although Labat (1742: vol. 4, 26) does mention large boiling houses with two. There were usually five or six cauldrons in a battery of which the first was the largest, about the size of the receiver, and each successive cauldron was smaller than the one before it. The battery was designed so that greater heat could be applied to the smaller cauldrons than to the larger ones (Fig. 5.4). In the battery the cane juice was cleared of non-saccharine impurities through a simple form of clarification, and then the water content was evaporated to the point at which sugar crystals would form in the remaining liquid. The contemporary authorities (Ligon 1657; Antonil [1711] 1968; Labat 1742; La Couture 1790) differ a little in their descriptions of the techniques, and what follows here is a generalized account. The juice was clarified in the first two cauldrons – indeed, the first of these was known to the English as the clarifier – and boiled rapidly to the point of crystallization in the remaining cauldrons. The heat was applied gently to the clarifier to heat but not boil the juice. Impurities rose to the surface where they formed a scum which the slaves skimmed off before transferring the juice to the second cauldron. The juice was then brought to the boil and a temper of alkalies added – lye, lime, ash and bone – to precipitate the remaining impurities. The quantity and kind of alkali appears to have been left to the discretion of the person in charge, the sugar-master, and in some boiling houses alkalies were added to the juice in the clarifier. The liquid boiled more vigorously in the third cauldron which was set over a stronger heat than the first two and threw off the remaining impurities. The juice was by this stage much reduced in quantity and in Brazil at least was referred to as syrup. Boiling continued in the fourth cauldron for a longer period of time than in any of the others until no more than half the original amount remained. The work then required a good deal of attention and slaves had to stir frequently the viscous syrup to prevent burning and the formation of caramel. In the final cauldron the syrup was reduced to a magma. It was removed from the heat when the sugar-master judged it would separate into crystals and molasses. This was the "strike point" and skill in determining when it had been reached could make or mar a batch of sugar. After the strike, the magma was removed to another container to cool for a few hours before the next stage in manufacturing began. The sources are vague on the time required to clarify the juice and reduce it to the strike point but three to four hours is probably a reasonable estimate. The battery was kept in continuous operation around the clock. As soon as a cauldron was emptied, it was immediately cleaned and refilled.

The boiling house employed surprisingly little labor compared to the other aspects of plantation life. Pomet (1694), for instance, shows only two slaves at work in a small boiling house in the French Caribbean (Plate 7) which suggests that four slaves or perhaps six if there were three shifts would suffice to keep the house in operation twenty-four hours a day. Out of a plantation complement of 120 slaves, Labat (1742: vol. 4, 174) assigned only six to care for the boiling juice and three to stoke the furnaces. The most detailed account of the sugar industry in colonial Brazil gives the impression of heavier staffing in the boiling houses. Its author, Father Antonil (1968: 198–9, 208–11), reported that six stokers and sixteen slaves tended the battery. There was a slave at each cauldron. Those stationed at the first two cauldrons were relieved more frequently than those attending the smaller ones, the work being more exhausting as there was more liquid to ladle and impurities to skim. The sugar-master was usually in attendance during the day and was relieved by an overseer at night. There was indeed much inconclusive discussion in the industry about the wisdom of continuing to boil through the night when supervision was less satisfactory (Barrett 1965: 161). The contemporary accounts have left opposing views of what it was like to work in boiling houses. Many an illustration shows them to have been places of order and calm with two or three slaves poised in elegant movement, ladle or skimmer to hand. The reality surely was different, if not quite so bad as the famous comparison made by the great Portuguese Jesuit Father Antônio Vieira who in a sermon preached in Bahia in 1633 likened the sight of men toiling at the cauldrons amidst eternal flames and smoke to a vision of souls in Hell.

The final stages in the manufacture of sugar took place in a building known either as the curing or the purging house. It was usually a large one-storey building with space for work and storage. Much of the space was taken up with rows of racks designed to hold the barrels and earthenware cones in which the sugar "cured." These containers varied in size from place to place, a fact which can complicate discussion of the amount of sugar actually exported from a given colony. The cones, for instance, could contain as little as 25 pounds or as much as 60 pounds of uncured sugar. There was a hole in the base of each barrel and in the apex of each cone – the cones were suspended in the racks apex down – to allow the molasses to drain into troughs that led to a storage tank. In some curing houses, each cone was inverted directly into and rested on the neck of a receptacle for the molasses.

The plantations of tropical America made several different types of granular sugar and each had its own name. The names for some of these types passed out of use over the years and new ones were introduced. Language has further complicated the matter: the English, French and Dutch borrowed names from the Spanish and Portuguese but also devised others of their own. The result is a rather confusing nomenclature, but two important types of sugar – muscovado and clayed or white – carried the same name throughout colonial America.

The muscovado was the easiest to make of all the sugars. Slaves filled barrels or cones with the magma, now cool, from the boiling house and set them on the racks for the molasses to drain, and after three or four weeks it was ready for shipment. Muscovado was a sticky brown-colored sugar that usually continued to ooze molasses on its way across the Atlantic. The production of clayed sugar differed from muscovado in that it was cured only in the cones and water was used to leach out the molasses. After the cones had stood for some time, slaves tapped the top of the sugar gently to compact it a little and then laid on caps of damp clay. Water from the clay percolating through the sugar washed out the molasses. This molasses was different in quality from that which dripped from muscovado and was therefore kept separately. Occasionally, the overseer would order a second layer of clay when the first had dried out. Within a few days, the slaves removed the sugar now shaped and compacted into loaves and placed them in the sun to dry. The sugar graded from white at the bottom of the loaf where it had been next to the clay, to brown or even muscovado at the apex. The technique of claying to obtain white sugar was an old one, brought over to the Americas by the Spanish and Portuguese. As noted in chapter 3, archaeological evidence suggest that Moroccan sugar manufacturers used the technique as did the Canarian planters (Nichols 1963: 111) and Andalusian legal documents from the 1560s refer to it (Ruiz 1964). Its origins may lie much earlier in the history of the sugar industry. There is, nevertheless, one explanation in the literature of how the properties of claying were discovered: workers in a curing house, presumably in a region of clay soils, noticed a trail of white foot prints where a chicken had walked on muscovado sugar. Although this story may well be apocryphal, many authorities (Anonymous 1956: 165–6; Sloane 1707: lxi) have repeated it over the centuries since what might well be its first appearance in western literature in Hernandez (1615: 58).

Sugar producers were determined to classify. Antonil (1968: 264–5), for instance, identified three grades of white or clayed sugar and, in a somewhat similar exercise, Mexicans assigned names according to whether the sugar came from the surface or the middle of the loaf (Barrett 1970: 60). There was, however, a basic distinction between muscovado and clayed sugars that were made directly from cane juice and those made by boiling and crystallizing the molasses that drained from the barrels and cones. The Brazilians attempted to avoid confusion by adding the adjectives *macho* and *batido* to distinguish between sugars from the first and second boilings as in *muscovado macho* (first boiling) and *muscovado batido* (second boiling). They did not consider the molasses that drained from the muscovado as suitable for making sugar and used only the molasses that drained from the clayed sugar for their *batido* sugars. The hacienda of the Marqueses del Valle, in Mexico, made sugar from both types of molasses: molasses from muscovado gave *azúcar de espumas* and from clayed sugar *panelas*. Some producers even had a third round, using the molasses that drained from sugars made at the second boiling (Antonil 1968: 264–7; Barrett 1970: 60–2). The early Barbadians appear to have had a simpler view of these matters, using, according to Ligon (1657: 91), "peneles," a misspelt borrowing, for all sugar made from molasses.

Refined sugars constituted yet a third group and are particularly associated with the French. Labat (1742: vol. 4, 121–50) described them in some detail, distinguishing between those made from muscovado and those made from clayed sugar. Essentially, the method of manufacture was the same as that in the refineries of Europe. The refiners dissolved sugar in water, repeated the familiar process of boiling and clarifying, but also strained the magma through cloth before setting it in cones or other molds to cure and dry. They took care to use the best clay, and the result was a white and pure sugar. Refined sugars graced the tables of wealthy planters; soldiers, sailors and officials took it home to France as souvenirs of their stay in the Indies, and some was exported.

Circumstances largely beyond their control determined the type of sugar the colonists produced. In the manufacture of sugar, there was a basic and inevitable conflict of interest between the colonies and the mother countries. The manufacture of sugar created employment and the further the colonists took the process of manufacturing towards refined sugar the less work they left for refiners at home. Each step, moreover, in the refining of sugar reduced the bulk of the end product thereby reducing the cargoes and injuring the interests of merchant marines. Already in the Mediterranean industry, these same factors had led the importers and consumers to acquire as much as possible of the manufacturing process from the growers and now they operated to control the activities of sugar-masters in the Caribbean. Politics had the support of geography in that the shortage of fuel in the Caribbean as well as the risk of damage to the sugar during the transatlantic voyage worked in favor of the European refineries. An Act of the English parliament in 1685 that raised high import taxes on refined sugar effectively put an end to refining in the English colonies. The French domestic refining industry was much less well-developed than the English one and for some years the French government actively encouraged refining in its new colonies so that France would not have to import sugar from refineries in northern Europe, but once refiners had established themselves in France the government moved quickly to stop the refining of sugar in its colonies. Muscovado sugar, from the first boiling, represented the necessary minimum amount of processing before export and few English plantations did more than this except on Barbados where claying sugar remained a common practice. The French planters continued to clay much of their production, perhaps because its smaller bulk and easier freighting was important to a nation that was not always able to protect its merchant marine in times of war (Pares 1960: 23–4). Muscovado and clayed sugar constituted by far the greater part of American sugar exports with sugars from the second boiling meeting a modest demand for better quality sugar within the colonies.

Two important by-products of the manufacture of sugar which entered international trade were molasses and rum. In England at least, molasses came to be a valued item in its own right, finding a place at first in poorer households as a sweetener and eventually becoming an ingredient in the manufacture of some kinds of biscuits. New Englanders, too, found ways to use molasses in their cooking, incorporating it even in such traditional dishes as baked beans. But molasses, together with the skimmings from the cauldrons, was the raw material for making rum and distilling was ubiquitous in the sugar colonies. Rum was of course drunk in the colonies but much was exported. The French, Spanish and Portuguese colonies, unlike the English, faced the problem that rum competed with brandy at home. The protection given to brandy in France made rum almost unsaleable there so that French planters had little market for their molasses and sold it at a very low price, often to New Englanders (Parry and Sherlock 1956: 129). The English planters broadened their market when in 1775 they persuaded the Navy to issue seamen with rum instead of brandy.

The merchandising of sugar

The trade of the plantations had to be organized and financed which created not only employment for seamen but extensive networks of commercial middlemen who transported and sold the plantation products and brought to the plantations slaves, food and consumer goods. In the early years of the industry, much of the capital for establishing plantations and many of those involved in the sugar trade were Genoese. They were joined in these activities by Sephardic Jews and "New Christians" from Portugal who were particularly active in Brazil until the mid-seventeenth century and afterwards in the Dutch and English Caribbean as well as in Spanish America. Dutch capital had been important in the mid-seventeenth-century Caribbean but English, French, Danish and even Swedish capital made its way into the industry. Each country, in mercantilist fashion, tried to restrict its colonial trade to its own nationals but such policies were often unenforceable and unrealistic. With its markets in Europe and North America, with food supplies coming from Europe, North and South America, with its main source of labor in Africa and with several European powers participating, viewed as a whole the eighteenth-century sugar industry was a multinational undertaking.

The organization of this trade varied a little from one colonial empire to another but there were many common characteristics. Merchants based in the great port cities of western Europe, or to a much lesser extent in American ports, financed the trade. They owned the ships, bought and sold the cargoes and acted as bankers to planters. The advantage of a European base for a merchant was proximity to markets, access to capital and contacts with a wide sphere of activity; the advantage of an American base was knowledge of the local scene, of which planters were creditworthy. The European-based merchants required trustworthy American representatives. Some in fact commissioned the captains of their ships, whom they knew well and saw at the end of each voyage, to buy and sell in the colonies but such an arrangement was not ideal as transaction of the business could take time while the ship lay idle in port with a crew to be paid and the captain as transient in a colony was unlikely to come to know well the financial background of the local planters. Only a resident was likely to acquire such commercial knowledge. An alternative strategy, therefore, was for a merchant to place an agent in each port where he traded who would provide advice on loans and look after investments, have a cargo ready for the ships when they called and sell the goods that they had brought. Many a merchant chose brothers, sons and nephews as agents, but as some found to their cost, kinship was not a guarantee of a mutually profitable business relationship.

There arose during the eighteenth century an alternative system of trade to the merchant as entrepreneur that became dominant in the English colonies. The planter consigned his produce to an agent or factor in England who paid the freights and duties, sold the produce, taking a commission, of course, and bought and shipped to the planter basic supplies and luxury goods. The factor acted as banker, keeping an account of a planter's receipts and expenses, accepting bills of exchange the planter might send in settlement of such major purchases as new slaves and providing credit when necessary, which it often was. Many an English planter became heavily in debt to his factor: a taste for luxury and excessively generous encumberances on estates in favor of heirs contributed to the decay of many a West Indian fortune. With the rise of the new system of trade, the population of resident merchants in the English Caribbean began to decline. Another cause of the decline was the lack of a market for imported consumer goods in a society where most people were slaves. Some merchants did survive, of course, particularly in Kingston, Jamaica, which enjoyed considerable trade with Spanish America, and in Barbados, which was a focal point in the slave trade. Trade in the French island did not undergo this change:

although some planters especially in St. Domingue shipped sugar to France on their own account the entrepreneurial merchant continued to play a major role.

Formal joint stock or government trading companies active in many other branches of European colonization and trade had little role in the sugar trade. The English sugar planters successfully resisted various proposals for companies in the later 1600s, and the French only briefly maintained a West India Company between 1664 and 1674. The purpose of the Pernambuco and Paraiba Company (1759–80) was to revive the depressed state of the economy of that part of northeast Brazil. Companies were more in evidence in the slave trade. The Royal African Company not only supplied slaves but advanced credit to planters and companies bidded for *asientos*, the contracts to supply slaves to Spanish America.

Within the local and regional markets of mainland Spanish America, the quantities of sugar traded and the amounts of capital required to fund the business were much more modest than in the international market. Much of the capital invested in the sugar industry was probably raised within the colonies from profits in mining and trade, and when a landowner needed a loan he could approach other landowners, the merchants or the church. Although the regional industries of Peru and Mexico did not generate the movements of men, money and materials on nearly the same scale as the international producers, they were significant centers of colonial agricultural enterprise (Pares 1960: 26–37; Sheridan 1974: 282–8; Lockhart and Schwartz 1983: 221–5).

The world that sugar made

Already by the beginning of the eighteenth century, the colonies that produced sugar for the international market had created a distinctive cultural and economic region in tropical America, and this region continued to expand throughout the century in response to market forces. By the late 1700s, it embraced coastal Rio de Janeiro, coastal northeast Brazil, parts of the Guiana coast, Barbados and many of the Lesser Antilles, Jamaica, St. Domingue and Cuba, and it was still expanding. There were "sugar frontiers" where new plantations were being hacked out of the forest in Brazil, in the so-called "ceded" islands – St. Lucia, St. Vincent, Grenada, Dominica – that Britain took from France in 1763, and most impressively of all in Cuba. The fundamental characteristic of the region was, of course, the dominance in the economy of the sugar industry but the dependence of the region on slave labor and the large concentrations of population also served to set it apart from the rest of colonial America. There were variations in densities of population and in the management of resources among the sugar colonies but the characteristics they had in common far outweighed the differences and ensured the region a unique cultural and economic identity.

The sugar industry supported the densest concentrations of rural population in the Americas. Over much of the Americas by the end of the eighteenth century there were no more than one or two persons per square kilometer. Along the eastern seaboard of North America, population densities as high as twenty per square kilometer were rare and found over small areas in a few favored localities such as coastal Pennsylvania (Lemon 1972: 62), New Jersey (Wacker 1975: 138) and Massachusetts (McManis 1975: 72). In Guanajuato, which was one of the most densely populated areas of New Spain, there were in the years around 1800 some thirty-six inhabitants per square kilometer (Sánchez-Albornoz 1984: 34). By contrast, Barbados held towards the end of the century approximately 78,000 people or 181 per square kilometer. Virtually the entire area of Barbados was cultivated which in part accounts for the exceptional density of people, even by sugar island standards, but nevertheless densities were only a little less, 142 and 129 inhabitants per square kilometer respectively, on St. Kitts and Nevis which have rugged uncultivatable interiors. The colonies that produced the most sugar had the most people, 280,000 in Jamaica and 535,000 in St. Domingue, although the densities there were less impressive – an average of nineteen per square kilometer on St. Domingue and twenty-five on Jamaica – but these figures underrepresent the densities in the cultivated areas as both colonies contained extensive, sparsely populated mountain ranges that were outside the sugar economy. The contrast between the sugar colonies and the neglected Spanish Caribbean colonies was most marked. Santo Domingo, Puerto Rico and, until the mid-years of the century, Cuba were largely uncultivated and inhabited by only a few thousand settlers. Trinidad in the 1780s had a population of 1,000. As an island was drawn into the international sugar economy, so the population grew, a process that can be clearly seen in the demography of St. Domingue and Cuba (Table 5.1).

There was also in Brazil a close correlation between population and the sugar industry although the inadequacies of the data mean that any figures on densities of population are very unreliable. In the 1770s, the cane-growing captaincies of the northeast had a population of some 580,000 (Alden 1963: 191). Most of these people would have been on the coast but a certain percentage was scattered through the vast interior. Some of the 215,000 people in the captaincy of Rio de Janeiro lived from the sugar industry but just how many did so is rather difficult to say. In comparison to the rest of Brazil and indeed to most of the Americas, the sugar lands of Brazil carried large populations, but the densities were less than in the Caribbean: Jamaica, for instance, at the end of the eighteenth century had the same number of inhabitants as the entire captaincy of Pernambuco.

Table 5.1 *Population of the West Indian sugar colonies during the eighteenth century*

	Antigua		St. Kitts		Nevis	
	White	Black	White	Black	White	Black
1678	2,308	2,172	1,897	1,436	3,521	3,860
1708	2,909	12,943	1,670	3,258	1,104	3,676
1720	3,652	19,186	2,800	7,321	1,343	5,689
1745	3,538	27,892	2,377	19,174	857	6,511
1756	3,435	31,428	2,783	21,891	1,118	8,380
1775	2,590	37,808	1,900	23,462	1,000	11,000

	Barbados		Jamaica	
	White	Black	White	Black
1700	15,000	40,000	7,000	40,000
1713	16,000	45,000	7,000	55,000
1722				80,000
1734		46,400		
1754				130,000
1757	16,772	63,000		
1775				190,000
1789				250,000
1792		64,300		

	Martinique			Guadeloupe		
	White	Black	Free colored	White	Black	Free colored
1686		8,000			4,600	
1720		36,000		5,000	17,000	
1767	12,450			10,000		
1789	10,634	83,400	5,000	13,712	85,500	3,000

	St. Domingue			Cuba		
	White	Black	Free colored	White	Black	Free colored
1681	4,000					
1686		3,400				
1713	5,000					
1720		47,000				
1730	10,000	80,000				
1775	20,000			95,419	44,336	30,615
1789	30,831	452,000	24,000			
1791		480,000				
1792					84,590	

Total population, 1792: 272,301

Sources: Antigua, St. Kitts and Nevis: Sheridan 1974: 150. Barbados and Jamaica: Dunn 1972: 312; Curtin 1969: 59. Martinique, Guadeloupe, St. Domingue and Cuba: Curtin 1969: 78; Pluchon 1982: 114, 170, 175; Edwards 1793–1801: vol. 3, 159–64; Thomas 1971: 65, 169.

A few more figures will reinforce the point that the sugar industry and high densities of population were closely correlated in eighteenth-century America. During the last years of the century, the total population of the Caribbean sugar colonies was approximately 1.5 million and if allowance is made for Brazil, perhaps 2 million people lived in that region of tropical America producing sugar for the international market. In all of Spanish America, from California to Tierra del Fuego, there were only about 13.5 million inhabitants (Sánchez-Albornoz 1984: 34) while according to the census of 1790, there were but 4 million people in the United States.

In addition to numbers and density, the high proportion of African slaves in the population was another distinguishing feature of sugar island demography. As the sugar revolution took hold in one colony after the other, so the importance of slaves in the total population grew. In Cuba, where this revolution was still in progress, slaves accounted for only 20% of the population in 1775 and 25% in 1792, but in those colonies with a long tradition of sugar cane cultivation, slaves made up a higher proportion of the total. In Bahia and Pernambuco, although the data do not permit a precise calculation, towards the end of the century the slaves were in the majority (Alden 1963: 197) but in some Caribbean islands they amounted to 85% or more of the total (Table 5.1). With the exception of Cuba, slaves formed a higher proportion of the population of the sugar colonies than in another major region of slave labor: the American South. There, according to Potter (1984: 138), writing of the South as a whole, "the Negro [all slave?] proportion was always over 20 percent after 1700 and always over 30 percent after 1740," but did not exceed 40%. South Carolina recorded the greatest percentage – 70% – in 1720 but this peak gradually leveled off to less than 50% by the end of the century.

The population of the sugar colonies was not only one of immigrants, but continuing immigration was necessary to sustain it. A naturally decreasing slave population was characteristic of northeast Brazil and the Caribbean sugar colonies (Curtin 1969: 28–9). An imbalance in the sexes of the arriving slaves is a part of the explanation, but the relentless hard labor combined with poor food, accidents and disease ensured a short life expectancy on the sugar plantations. The figures, although approximate and still under discussion, of the numbers of slaves sugar colonies imported are revealing of the harshness of slave life. Barbados, for instance, had a slave population of about 40,000 in 1700 and after importing 263,000 slaves during the next hundred years, some of whom admittedly were reexported, its slave population in 1800 was no more than some 66,000. The growth in the slave population of Jamaica from 45,000 in 1702 to 324,000 in 1808 required the import of 659,000 slaves; St. Domingue imported 864,000 slaves to raise its slave population from 4,000 in 1680 to 480,000 in 1791. A short life expectancy was the norm for the slaves on the sugar plantations and this characteristic is another distinguishing feature of the cane-growing region. On the eastern seaboard of North America the statistics permit the inference that there the conditions of life and labor for slaves were much better. By 1790, the thirteen colonies had imported 275,000 slaves, but the slave population in that year stood at approximately 700,000. Over the entire period of the slave trade, some 4 million slaves were brought to the Caribbean, another 530,000 to the Guianas and of the 3.6 million sent to Brazil a good number went to the sugar plantations (Curtin 1969: 53, 55, 59, 72-3, 78-9, 88-9).

The white population of French and English colonies also required immigration to sustain it, although this was perhaps less true of Cuba and Brazil. White society in the English and French Caribbean consisted of an elite of wealthy planters and senior government officials who lorded over a middle class of military officers, clergy, agents and merchants and skilled craftsmen, and a lower class of "poor white" servants. Their white color gave this disparate group a common identity, fear of the slaves a common interest and making money a common motive, but the narrow rounds of this society could offer little in the way of security, education or entertainment to shield its members from the exploitation on which it was based. Those who could escape did so before death from gun, knife or one disease or another overtook them, to live at "home" on revenues from their plantations or on the savings from trade. Administrators and the military were usually temporary residents who served terms of duty, leaving the "poor whites" as the least likely of the white population to find an escape from the islands. Without the arrival of new servants and of young men embarking on careers in administration, in the military or in trade, the white population of the British and French islands would have dwindled to far lower levels than was the case. In contrast, the white populations of Cuba and Brazil were more committed to the colonial life: the absentee proprietor was comparatively rare and while it was not unusual for the sons of the well-to-do to go to Europe to study, many of them returned within a few years. Moreover, there was a flow of immigrants from Spain and Portugal whether driven by necessity from poverty stricken villages or attracted by career opportunities.

The European immigration to the sugar colonies was largely male and so many naturally turned to the black women for wives and mistresses with the result that there emerged a segment of the population that was of mixed racial descent. "People of color" were discriminated against in the French and British West Indies, but their lot was better in Brazil where in fact many were accepted as being white if they had wealth or good social connections. Given the problems of definition and the inadequacies of the demographic data for the colonies – the statistics from the English islands do not record them – it is difficult now to know what proportion these people formed of the total population. The data from the French islands suggest that at the end of the eighteenth century they amounted to about 5% of the population, rather more in Cuba.

There is an old Portuguese saying of Brazil that it was a hell for blacks, a purgatory for whites and a paradise for mulattoes and while the free people of color in Brazil and other sugar colonies would certainly have jibbed at the use of the word paradise to describe their lot and only the better-off whites could attain if not salvation at least a return passage to Europe, there can be little argument about the place of the blacks. Their response was protest which took the several forms of suicide, vandalism, attacks on overseers and owners, rebellion and flight. This last alternative led to some significant and lasting modifications in the cultural geography of the sugar region.

In all the sugar colonies slaves tried to escape and the term for a runaway – a maroon – was in use as early as the 1530s (Price 1979: 1–2). Success in running away was in good measure at least a reflection of geographical circumstance. Small islands, densely populated and intensively exploited, offered little in the way of a refuge for the runaways and escape was difficult but in the mountainous and forested interiors of Jamaica, St. Domingue and Cuba they could evade the authorities and establish their own communities. Maroon settlements, too, became common along the inland margins of the plantation areas of the Guianas and Brazil. Ironically, perhaps, these settlements that were the creation of runaway slaves were nevertheless parasitical on the plantations. The maroons stole tools, clothing and other goods from the plantations; they sought new recruits there, particularly women, but they also looked to the plantations as a market for the game they hunted and the food crops they grew. For the colonial authorities the maroons were seen not only as a threat to law and order but as an unwelcome example to the slaves, and they organized military campaigns against them, not always with success. The most famous maroon community of northeast Brazil, Palmares, survived from about 1605 to 1694. The English were not able to subdue the maroons of Jamaica and at the end of the First Maroon War, in 1740, signed a treaty which guaranteed them their freedom, rights to land and permission to sell crops in the island markets. For their part, the maroons agreed in what can be regarded as a rather unworthy clause to round up and return to their owners all future runaways. Maroons were part of the frontier of plantation society, occupying lands that were either unsuited to sugar cane cultivation or that in time would be absorbed by an expanding sugar industry. They were smallholders, peasant cultivators, in at the origins of West Indian peasant society. Their legacy extends to the villages they founded, a number of which survive, some now grown into towns.

Plantation society was a rural society in which the plantation was not only the unit of production but also the basic functional unit in the settlement pattern. Each nucleus of population – slaves, overseers, manager or master – was separated by cane-pieces, pasture and forest from the next. Most Caribbean islands had only one settlement that deserved the description of town and it was usually the administrative center and port. The largest towns in the region at the end of the eighteenth century were Bahia and Havana, both with about 50,000 inhabitants, which made them large by American standards of the time. The sugar economy does not entirely account for their size: Bahia had been until 1763 the capital of Brazil, and Havana was not only an administrative center but a strategic port and garrison. The largest town in the English colonies was Kingston, Jamaica, with 5,000 inhabitants in 1700 and 25,000 in 1790 (Clarke 1975: 6).

The main lineaments of the geography of the sugar region of tropical America can now be summarized. It was insular and coastal, because of the industry's need for easy access by ocean transport to European and, later, North American markets. It was dominated, of course, by the sugar industry although not entirely to the exclusion of other forms of land use. Tobacco, indigo, cotton and coffee were also grown in some parts of the region. While the production of sugar made heavy demands on the natural resources of the region and led to the clearing of much of the forest cover and in places to soil exhaustion, the planters were able to adapt their land use strategies to overcome these difficulties and so preserve their investments in land, labor and equipment. The adoption of conservationist innovations did, however, make the cultivation of sugar cane more labor-intensive. Because of the concentration of land and labor on production of the cash crop, the sugar region was the only major agricultural region of colonial America that could not feed its own population. Even the most basic staple foods had to be imported. Europe at first was the supplier, then the eastern seaboard of North America became another source and later still temperate South America. Nor did the sugar colonies manage to raise sufficient livestock. There was a trade in cattle to work mills and pull carts of the sugar colonies from the neglected Spanish islands of the Caribbean as well as from the interior of northeast Brazil. Finally, this highly commercial agricultural economy supported the greatest densities of population in colonial America, a population, moreover, that required a continuing inflow of immigrants, slave and free, to maintain the numbers.

However viewed, the society of the sugar region was a vulnerable society. First and foremost, it depended on the successful marketing of one cash crop – sugar cane – and therefore its financial viability could be threatened by competition from other producers of sugar cane or by the appearance of an alternative sweetener. It depended also on slavery, and while it had always lived with the fear of a successful slave revolt, by the end of the eighteenth century the abolition of slavery by the metropolitan governments had become a distinct possibility. The third looming source of disruption was technological change. There had been no significant development in this field since the introduction of the three-roller mill and Jamaica train but the use of steam power in Europe and the growing size of factories there suggested that the introduction of a new generation of sugar mills, larger and more efficient, and which in turn would lead to the reorganization of the landholdings, could not be far off. Indeed, already, by the end of the century, the first experimental use of steam to power sugar mills had begun in Jamaica and Cuba (Fraginals 1976: 37–8).

With the benefit of hindsight, we now know, of course, that the sugar cane industry during the last years of the eighteenth century was at the end of a continuous chain of development that is traceable back to the medieval Mediterranean. From the time of the Arab agricultural revolution, the increasing production of cane sugar had led gradually in the western world not only to the replacement of other sweeteners but to a reduction in the price of sugar so that it had become affordable even by the poor; the association between the sugar industry and slavery, already discernible on Crete and Cyprus, had strengthened into complete dependence while the industry had grown in importance as an instrument of colonial expansion to become in the West Indies and Brazil the fulcrum of the Atlantic commercial empires. In the nineteenth century, the slow evolution of a thousand years was to give way to rapid transformation.

6

The innovations of a long nineteenth century: 1790–1914

The world of the nineteenth century was far less kind to the American sugar colonies than the one they had known before. As Europe industrialized and looked to America to supply raw materials such as cotton and timber for its factories and cheap wheat and beef to feed the factory workers, the flow of trade and investment changed, to the detriment of the sugar colonies. Moreover, nineteenth-century patterns of immigration dealt these colonies a second blow: people moved from Europe to the temperate American lands and the growth of population in the United States and Canada, in Argentina and southern Brazil soon provided richer markets for manufactured goods than the Caribbean islands. The changing world order was accompanied by a change in ideology: mercantilism gave way to liberalism, and free traders in Britain argued the virtues of the freedom to buy where goods were cheapest and sell where the prices were highest as they sought to benefit from the fortunate fact that their country was first into the industrial race. In this new world, the sugar colonies soon lost their pivotal role in Atlantic commerce and with it the protection of the imperial powers.

Disregard for the old sugar colonies was made all the more easy with the discovery that sugar beet could yield a sweetener indistinguishable from cane sugar. The first commercial beet sugar appeared on the market in the last years of the eighteenth century and beet eventually was to capture a large share of the market for sweeteners. Competition for the American industry came also from new producers of cane sugar, the result of the extension of colonialism into Asia, Africa and the Pacific where cane sugar industries in Bengal, Mauritius, Réunion, Java, Fiji, Natal, Queensland and Hawaii began to contribute to the international sugar trade. A major consequence of the entry onto the scene of the beet industry along with the expansion of the cane sugar industry was a great decline in the price of sugar. Prices might have fallen even more dramatically had not the growth in the world's population and the extraordinary increase in the per capita consumption of sugar in the industrializing countries shored up the demand.

The nineteenth century was therefore a period of crisis for the American sugar colonies in which planters had to meet the pressures of competition and declining prices, but they had also to adapt to other forces of change that by the end of the century were to have completely transformed the industry. Perhaps fittingly, the herald of the end of the old order was the beginning of the abolition of slavery.

Slavery had always had its critics. Not until the eighteenth century, however, did criticism of slavery become so general in west European countries as to make abolitionism into a serious political force. The great figures of the Enlightenment had debated the legal and moral basis of slavery, and in so doing had informed an increasingly literate public of the issues. A humanitarian critique came to the fore, strongly promulgated by religious groups such as the Quakers, and it is indeed this humanitarianism that has become fixed in the British popular mind as the driving force of the campaign not only to free the slaves in the British colonies, but to end slavery and the slave trade anywhere they were to be found. A second line of criticism was economic in nature. The French physiocrats and Adam Smith argued that enlightened self-interest had an important role in the generation of wealth, and consequently workers would work more productively if they benefited from their own labor. As slaves clearly did not benefit from their own labor, they did not put out much effort, and indeed would find rewards in thwarting the interests of their owners. The public good was best served, so the argument ran, if the enlightened self-interest of both landowners and agricultural workers coincided and therefore the route to greater wealth – and cheaper sugar – lay in the freeing of the slaves. These liberal arguments were extremely attractive to European refiners who wished to import sugar for as low a price as possible as well as to industrialists for whom cheap sugar and cheap food meant they could pay low wages.

This strengthening alliance between the humanitarians and Liberals along with the decline in the importance of the West Indies to the economy of Europe, the growth of the free trade movement, and the entry into the political vocabulary of such slogans as Liberty, Equality and Fraternity made the defense of slavery a difficult undertaking. It was no longer plausible in the late eighteenth century to try to defend slavery on the grounds that it was a means of bringing pagans to Christianity, and indeed all that the owners could manage in their own support was a not very convincing plea of necessity: that without slaves, the plantation economy would collapse because of the lack of labor. Translated, this meant they wanted the assurance of a cheap supply of labor after abolition, a concern for which governments did show some sympathy. Planters, of course, were also concerned about compensation for the property they would lose by abolition. Although planter political influence waned rapidly in northern Europe it remained influential in Spain and the newly independent countries of Latin America so it took a long time to eradicate slavery form the American sugar plantations.

The disruption of the old order in the sugar industry and the beginning of the long transition to a new one began with the slave revolt in St. Domingue in 1791. The revolt, eventually, was successful in that it rid the colony of slavery and brought it within a few years to political independence under the name of Haiti, but it also led to the destruction of the sugar industry as well as of other types of commercial agriculture, to the murder or exile of virtually all members of the planter class and to decades of civil war. It was not an example that could give any comfort to planters watching the approach of abolition in their own societies. Britain's participation in the slave trade ceased as a result of an Act of parliament in 1807, but slavery continued in its West Indian colonies until 1834. Even then, the former slaves still had to serve their former owners as "apprentices" in a move to protect the plantations from an immediate loss of labor and they only acquired legal freedom of movement in 1838 when London declared "apprenticeship" at an end. Slavery persisted until 1848 in the French colonies, 1886 in Cuba, and only when Brazil freed its slaves in 1888 was the connection between the sugar industry and slavery that had begun in the eastern Mediterranean finally severed.

The abolition of slavery was a rending political, emotional and financial issue that dominated political debate in all the plantation societies for decades. When it was achieved, it had two immediate and far-reaching geographical consequences in the cane-growing regions. It changed the settlement pattern as former slaves abandoned the old slave quarters to live in new villages, and where they also abandoned work on the plantations it created a labor problem. The search for a new source of labor became virtually world-wide in scope and lead to the migration of some hundreds of thousands of people who in their turn further diversified the ethnic mix in the population of the sugar colonies. In other words, the role of the sugar cane industry as a catalyst in the formation of the cultural geography of the tropical world continued during the nineteenth century.

Technological innovation was another catalyst in the transformation of the industry. Before 1800, the milling and manufacture of sugar in the Americas were governed very much by routine, and the only two innovations of any significance were the three-roller mill and the Jamaica train. After 1800, the scientific discoveries that made possible the industrial revolution in Europe were applied also to the sugar industry. They set in motion a train of innovations that led to ever larger and more efficient mills and factories which in their turn had consequences for both the spatial

The innovations of a long nineteenth century: 1790–1914 123

organization and financial structure of the industry: the larger mills and factories increased the optimum size of plantations and so caused a consolidation of properties, while the expense of assembling land and financing the new technology took the industry out of the control of individual planters and their merchant-agents into the world of large corporations.

Sugar cane – the plant – also became the subject of scientific experiments. The discovery towards the end of the century that sugar cane could produce fertile seeds combined with the new field of research in plant genetics proved an enormous boon to the industry. It became possible to breed new varieties of sugar cane to improve yields, resist disease, suit different edaphic and climatic conditions. There can be no doubt that cane breeding has contributed in a major way to the survival of the industry.

In a real sense, therefore, the nineteenth century marks a break with the past for the American sugar industry. The gradual evolution that had characterized the industry over the centuries from garden cultivation in the Levant to the large plantations of the West Indies gave way to a pace and scope of change that was revolutionary in comparison. Indeed, between 1790 and 1914, every aspect of the cultivation of the cane and the manufacture of sugar changed, with the notable exception of hand-harvesting. Some colonies for one reason or another were unable to deal with the problems and, like St. Domingue and Jamaica, fell into poverty and obscurity while new producers like Cuba and Puerto Rico rose to prominence.

The forces of change

The labor problem

At the end of the eighteenth century, all the cane-growing regions in the Americas that produced for the international market relied to a very great extent for their labor on African slavery. Unfortunately, it is not possible to give a precise figure for the numbers of slaves employed in the sugar industry at this time: the censuses of slave populations that survive are of uneven accuracy and, moreover, they seldom break down the information they provide into occupational groups. Slaves worked not only in the sugar industry but also in other branches of agriculture, in mining, as urban laborers and as household servants. Nevertheless, the distribution of the slave population shows a strong bias to those regions in which the cultivation of sugar cane was the most important economic activity. In 1790, there were about 2,500,000 slaves in the Caribbean, Central and South America of whom 1,250,000, or 50% of the total, worked in the Caribbean islands, about 750,000 were in Brazil and the remaining 500,000 were in Spanish central and south America. The Caribbean concentration was even greater than this outline suggests because a good percentage of the mainland Spanish American slaves were settled along the Caribbean coast of present-day Venezuela and Colombia. By contrast, there were only 10,000 slaves in Mexico and 40,000 in Peru (Alden 1984; Bowser 1984; Curtin 1969).

These figures illustrate the magnitude of the problem that abolition meant for the sugar planters. Not only was their investment in the slaves at stake but if they were to keep their plantations and mills in production they either had to find some means of persuading the former slaves to continue to work in the sugar industry or, if their blandishments failed to secure a sufficient labor supply, replace hundreds of thousands of former slaves with laborers from some new source. And even if a good number of former slaves did accept work on the plantations, the planters were well aware that an influx of laborers would be a means of keeping wages down. For a planter class that had relied on slavery for generations, the financial and logistical consequences of abolition were indeed daunting, and the slave revolt in St. Domingue made the future seem yet more problematical by raising the prospect of anarchy once the restraints on the slave population had been removed.

The availability of land had an important bearing on the severity of the labor problem that abolition caused. On some small Caribbean islands such as Barbados, St. Kitts and Antigua, where the sugar plantations occupied virtually all the land, the former slaves had no alternative but to continue to work in the sugar industry. In these islands there was in fact no shortage of labor after abolition, rather, as the nineteenth century wore on, there was a surplus and workers left these islands to find jobs elsewhere in the Caribbean and Central America (Richardson 1983). But where there was unoccupied land within easy reach of the plantations, the former slaves did have the option of abandoning the scenes of their servitude to make shift for themselves on an agricultural frontier. Such was the case, for instance, in both Jamaica and Trinidad. Sir Lionel Smith, governor of Jamaica, was being a little naive when he dismissed the complaints of the planters about the shortage of labor with the remark that the former slaves would present themselves for work if well enough paid (Thomas 1974: 10–11). There was a limit to what the planters could pay given the declining price of sugar, but the matter of pay was not the only issue: the strong yearning to be free was another. And many former slaves demonstrated their belief that freedom could be more nearly achieved on a small-holding in the hills than laboring for a wage in the old plantation fields even when no longer the legal chattels of the planter.

Faced with the necessity of replacing the slaves, the planters thought first and foremost of attracting emigrants from Europe. Slavery, they recognized, had discouraged settlement by Europeans who understandably did not wish to work alongside slaves but with the removal of this impediment some of that stream of emigrants flowing into North America and other New World temperate lands might be deflected to the sugar colonies. The planters also had political and racist motives in opting for this source of labor: European workers would not only increase the white proportion in the population but might also establish a middle class that would block any upward mobility on the part of the former slaves and their descendants. Another way to limit the options of the former slaves and so keep more of them on the plantations was to settle Europeans on the vacant lands. The Jamaican planters saw the European settlement of the cool highlands of the island in this light (Laurence 1971: 9). But the promoters of such schemes had the impossible task of convincing would-be immigrants that tropical colonies with a history of slavery and high risk of disease offered better long-term prospects for a comfortable life than, for instance, a farm in Ohio. They attracted only a trickle of settlers from northern Europe but had considerable success in Madeira. This former sugar colony had very little in the way of economic opportunity to offer its inhabitants and already by the early nineteenth century Madeirans had a long tradition of emigration to the Americas. They began to appear in the British West Indies during the 1830s and continued to emigrate there for fifty years, helped for part of this time by government assistance towards the cost of the passage. Their main destination was British Guiana which received between 1835 and 1881 some 30,000 Madeirans (Moore 1975: 3).

Another option both the English and French planters explored was the recruitment of free labor from Africa. They proposed a form of indentured labor whereby in return for an assisted passage to the islands and guaranteed paid work, the immigrant would remain for a fixed term on a sugar plantation. Sierra Leone and St. Helena appeared as particularly likely sources of supply. The British had founded Sierra Leone on the west coast of Africa in the late eighteenth century as a place to put the slaves freed in Britain, and during the nineteenth century the Royal Navy used both colonies as refuges for the slaves captured on the high seas in its attempt to suppress the Atlantic slave trade. Sierra Leone was a poor place with little prospect of commercial development while the limited size of St. Helena meant that it could not become the permanent home for all those Africans whom the Royal Navy put ashore. There was, therefore, some grounds for optimism that London would support an initiative to attract some of these liberated slaves to the sugar plantations. The Jamaican planters persisted for over twenty years, beginning in 1840 when they passed in their island Assembly an Act to Encourage Immigration and dispatched an official to London and Sierra Leone to plead their case, but they achieved disappointing results (Thomas 1974). Their difficulties were the same as those of the French planters in their attempts to recruit free African labor: the memories of the slave trade and slavery were too strong for Africans to accept labor on West Indian sugar plantations, and public opinion in France and Britain found this new traffic too reminiscent of slavery for governments to give it forceful encouragement. The statistics Schnakenbourg (1984: 92) has compiled show that indentured African labor made only a small contribution to solving the plantation labor problem: 17,000 to the French West Indies, 14,000 to British Guiana, 9,000 to Trinidad and 11,000 to Jamaica.

Asia and not Africa in the nineteenth century was to supply cheap labor for the European agricultural exploitation of the tropics. Trading empires in the East became also territorial ones, and the British in particular but also French and Dutch found themselves administering densely populated lands from which agricultural labor could be exported. The opening of China and Japan brought many more millions of people within the orbit of western interests. Steam ships greatly increased the speed and safety of ocean travel, making the transfer of large numbers of people from one continent to another much easier than before. Although the metropolitan governments were fearful that a traffic in laborers between Asia and the plantations might become a new form of slavery, they relented in the face of the importunings of their labor-short colonies. And so another great current of migration began that distributed workers not only to the Americas, but also to Africa, Fiji, Hawaii and to places within Asia such as Ceylon and Malaya where there was a demand. These emigrants went usually under the supervision of an imperial government, as indentured laborers, with passage paid, work guaranteed and with the promise of a passage home if they wished to return after their indenture was over.

Indians were to make a major contribution to solving the labour problem of the sugar plantations. In the 1820s, the French authorities on the Indian Ocean island of Réunion imported Indian laborers and soon Indians were also going to Mauritius. John Gladstone, an owner of sugar estates in British Guiana who had good political connections in England, received permission to employ Indians, and in January, 1836, he wrote to his agents to ask them to recruit 100 Indians for five to seven years. This experiment proved successful, and negotiations began between the West Indian authorities, the Colonial Office and the Indian government to establish Indian indentured labor in the West Indies on a firm basis. Finally, in 1844 the Indian government legalized emigration to Demerara (British Guiana), Trinidad and Jamaica. An emigration of indentured laborers then began to these colonies that continued until the Indian government put a stop to it in 1917, ostensibly on the grounds of reserving the manpower for India's own war effort (Tinker 1974: 61–115, 334–66). Martinique and Guadeloupe were also able to obtain Indian indentured laborers. They began to arrive in the 1850s, and the traffic was regularized in a convention signed by Britain and France in 1861. The system, however, did not work well in

Table 6.1 *Indian indentured laborers in the Caribbean, 1838–1917*

	Arrived	Returned	Percentage returned
British Guiana	238,909	75,547	31.62
Trindad	143,939	33,294	22.37
Jamaica	36,412	11,880	32.63
Martinique	25,519	5,000*	20.00*
Guadeloupe	45,000*	8,000*	18.00*
Surinam	34,304	11,559	33.70
Windwards	10,026	3,774	37.64

* Figures approximate.

the French Caribbean. Both British and Indian governments received from their representatives numerous complaints about the treatment accorded the laborers, and of particular concern was the difficulty the laborers experienced in returning home after the period of their indenture was over. Given this unsatisfactory situation, the Indian government could not encourage further emigration and it came to a halt in 1885 (Lasserre 1961: vol. 1, 305–15; Tinker 1974: 276–9). Statistics (Table 6.1). compiled by Laurence (1971: 57) show the importance of the Indian contribution to the labor force of the sugar plantations.

For those searching for cheap labor in the mid-years of the nineteenth century, China was an obvious place to turn. It had a weak government unable to protect its citizens, it was populous and poor, and the Taiping Rebellion (1849–65) had driven millions of destitute peasants into the ports of the south coast. Shortly after the signing of the "unequal" Treaty of Nanking in 1842, by which China opened its ports to foreign commerce, agents looking to recruit men for work overseas appeared.

In tropical America only two countries – Cuba and Peru – made significant use of Chinese labor on the sugar plantations. Both countries in the 1840s faced labor shortages. Cuba could not meet the labor demands of its expanding sugar economy either through the slave trade with Africa or by European immigration; in Peru planters were concerned about the consequences of the imminent abolition of slavery. The fact that both were still slave states – Peru until 1854 and Cuba until 1886 – was a major impediment to negotiating any agreement with Britain for Indian indentured labor. China seemed to offer a solution. The Cuban Council for Economic Development sanctioned the import of Chinese contract labor in 1846 (Turner 1974: 71). In 1849 Peru passed a new immigration law and as its purpose was to permit Chinese laborers it became popularly known as the "Chinese law" (Stewart 1951: 13). In both countries, the Chinese were to be indentured laborers bound to work for a given number of years for the person who had bought their contract, and to receive in return a minimal level of room and board and a pittance of a wage.

From the very beginning, this new trade in human beings was conducted in an appalling manner, with scant regard for the welfare of the laborers. They were taken on board in Canton and other ports along the south coast of China, including the Portuguese colony of Macao and, until 1855 when the authorities intervened, the British colony of Hong Kong. Few of them could have understood the nature of the indenture they had entered into, and besides deliberate deceit on the part of recruiters who were often paid on the basis of the numbers they brought to the ships, recruitment often involved other reprehensible practices such as kidnapping and the purchase of prisoners taken in local wars. Some of those on the American side who organized the trade, such as the rich Spanish-born, Cuban-based, merchant and planter Julián de Zulueta, had been involved in the African slave trade. The ships were overcrowded, the Chinese badly fed and brutally treated, and the death rate on the voyages high, about 15%. There were, unsurprisingly, given the circumstances of the trade, not only riots on board the ships but even mutinies that led to the death of crewmen. The affair of the *Nouvelle Penelope* in 1870 brought the true nature of this traffic into public prominence. This French ship had sailed from Macao for Peru with 310 Chinese laborers on board who a few days out to sea killed the captain and eight crew members and turned the ship back to China. One of the Chinese subsequently fled to Hong Kong from where the Chinese authorities tried to extradite him for his part in the mutiny. In the ensuing investigations, it emerged that the Chinese laborers had boarded the ship at Macao under armed guard and that at least 100 of them had no wish to go to Peru. The Hong Kong magistrate denied the extradition on the grounds that the plaintif and his companions were within their rights to resist with violence their enforced emigration, and he concluded from the evidence placed before him that "the commerce in coolies was a true slave trade" (Laurence 1971: 31–3; Stewart 1951: 48–51; Turner 1974).

A trade described officially in such a way could scarcely be allowed to continue. Portugal and Peru drew up a face-saving consular convention to govern contracts for services and shipping in each other's territory; the Chinese government stopped the trade in its own ports and in conjunction with the governments of Britain and the U.S.A. made strong representation to the Portuguese. Finally, on March 27, 1874, the governor of Macao banned the traffic (Stewart 1951: 52–3). By this date, 121,810 Chinese indentured laborers had been landed in Cuba and about 90,000 in Peru (Turner 1974: 81; Stewart 1851: 74). In Cuba, some of the Chinese worked in the cultivation of tobacco and coffee; in Peru, many worked at mining guano and in building railways; in both countries many served as domestic servants, but the great majority toiled on the sugar plantations where they were an important addition to the labor force. Their numbers in the agricultural work force inevitably soon declined as some moved to the towns and others died. The Cubans turned to Spain for new supplies of labor, and between 1882 and the start of the War of Independence in 1895 succeeded in attracting about 80,000 immigrants of whom many, however, managed to evade harsh plantation work to settle in the towns (Laurence 1971: 35). The Peruvians attempted during the remainder of the century to revive the emigration from China on a more humane basis, even trying to attract Chinese who had settled in California, but with no success (Stewart 1951: 160–223), and shortage of labor remained a constant concern of the Peruvian planters. At the very end of the century, they were able to tap a new source: Japan.

The social and economic changes set in motion by the Meiji Restoration of 1868 and the opening of Japan to world trade led to a rapid increase in population and a high rate of rural unemployment. By the late years of the century, the government was willing to permit emigration. This was managed by Japanese emigration companies, and the first American destinations of the Japanese were British Columbia and California. When Canada and the U.S.A. placed restrictions on the entry of Japanese, the emigration companies looked for other destinations and found one in Peru. The Japanese came under four-year contracts that clearly stipulated the terms of employment. They were better informed than their Chinese predecessors of their rights, and they enjoyed the protection of the Japanese legation in Peru. This migration of contract workers, organized by the emigration companies, began in 1899 and lasted until 1923, and brought 18,258 Japanese to Peru (Morimoto 1979: 11–55).

Members of other ethnic groups also found themselves transported across the world for labor on the sugar plantations of America, although in very small numbers compared to the Indians and Chinese. The French brought a few hundred Annamese to the West Indies and the Dutch some thousands of Javanese. A particularly sad instance of this international search for labor was the descent of Peruvians on the inhabitants of South Pacific islands who were carried off in conditions that verged on slavery. During seven months in 1862–3, 3,634 islanders were rounded up for work in Peru of whom some escaped and some died on the voyage, but 3,125 arrived in the port of Callao. The diplomatic representatives in Peru of Britain, France and Hawaii soon discovered this activity, apprised their own governments and protested to the Peruvians. The traffic was stopped, but because of deaths in Peru and during repatriation, only 148 reached home safely (Maude 1981). However, the Pacific islanders were to contribute in a major way to the labor force of the sugar industry of Queensland (see ch. 9).

The ultimate and the permanent solution to the labor problems of the plantations was found at home, in the population of the Americas which was growing rapidly during the nineteenth century through natural increase as well as through immigration. Plantation labor could never be made particularly attractive, but given the lack of other opportunities for work even a very modest cash income brought the very poor to the sugar industry if only for seasonal work. In the Caribbean, inter-island migration became common as people from densely populated islands such as Barbados and St. Kitts sought jobs in the expanding sugar industries of Trinidad, British Guiana and the Dominican Republic; Indians were induced to come down from the Andes to the plantations of coastal Peru and northwest Argentina; and in Brazil a migration began from the northeast of the country to the new sugar plantations in São Paulo.

The competition from beet sugar

Sugar cane was the first tropical crop to encounter competition from a plant that could be grown in temperate lands with all the advantages of proximity to the major markets. Refined beet sugar is indistinguishable from refined cane sugar and can be used for the same domestic and industrial purposes. During the nineteenth century, the expansion of beet sugar production had a depressing effect on sugar prices and since the 1880s, there has usually been a glut of sugar on the market except during the World Wars. The fact that sugar, whether beet or cane, can be produced cheaply in so many parts of the world has made the regulation of production and prices in order to guarantee at least a minimally acceptable return on investment for the growers very difficult to achieve.

Sugar beet is a common form of *Beta vulgaris* L., closely related to beetroot and mangels and like these beets can be eaten by humans or livestock. Sugar beet grows easily in Europe, particularly north and east of the Alps, and has come to be widely cultivated as a commercial crop in North America and Asia. Beets, usually biennials, accumulate sugar in their thickened roots. As early as the sixteenth century, some agriculturalists realized that red beet contained sugar, but it was not until 200 years later, in 1747, that a Berlin professor, Andreas Marggraf (1709–82) discovered a process for extracting sugar from red and white beets that yielded half an ounce of white sugar from half a pound of dried beet. Franz Carl Achard (1753–1821) continued Marggraf's work, and in 1799 asked the King of Prussia for assistance. Royal favor led to the building of a factory but it was not a success. This disappointment did not prevent the founding of several small factories in Silesia and in the neighborhood of Magdeburg. Other European countries became interested in these activities, especially France where Achard had sent reports on his experiments, and in the

Netherlands the Society for the Encouragement of Agriculture offered a prize for the best method of extracting sugar from native plants. The new industry, however, had not found by the end of the eighteenth century a firm commercial footing because as yet it could not match the cost of the imported cane sugar. It needed a protected market.

The Napoleonic Wars created such a market as England's blockade of Europe interrupted the supply of cane sugar to the continental states. Napoleon's government turned quickly to the one alternative source of sugar – beet – and the new industry entered a period of expansion. But there were financial and logistical problems that were difficult to overcome. Capital was not readily forthcoming to finance the factories; farmers often were unable to take their beet to the factories because of poor transportation; seed was in short supply; and many farmers did not know how best to cultivate the new crop that was being urged upon them. Understandably, the industry did not develop with the speed the authorities had wished, but nevertheless they remained optimistic, planning for 100,000 hectares of beet in 1812 when in 1811 only 6,785 had been planted in the entire Napoleonic empire. With the resumption of trade in cane sugar after 1815, the beet industry collapsed. The industry had been predominantly French, and it was only in France that any attempt was made to persist with the experiment (Slicher van Bath 1963: 276–7).

The revival of the beet industry began in the second quarter of the century at first in France, and then more generally across northern Europe in Belgium, the German states, Austria–Hungary and Russia (Fig. 6.1). One stimulus was the problems of the sugar cane industry caused by the abolition of slavery in the British and French colonies in 1833 and 1848 respectively, another was the realization on the part of landowners of the useful role sugar beet could play in the agricultural system. The cultivation of beet required deep plowing, and the generous use of fertilizer, and its inclusion in the rotation raised the yields of other crops, particularly grain. Moreover, beet leaves and roots could be used for cattle feed, thus increasing both stocking levels and the availability of manure. In brief, the cultivation of beet could substitute for fallow and was a means of intensifying land use. The role of beet in the agricultural economy began to increase markedly after the 1820s when the arrival of cheap grain from Russia depressed grain prices, forcing landowners to turn to cattle raising and to search for other cash crops. A final consideration in the explanation of this renewal of the industry was the fact that taxation policies of the beet-producing countries favored the beet industry: they levied tariffs on cane sugar but no country taxed beet sugar until France began to do so in 1837, and taxation became quite general in the 1840s. By this time, however, the sugar beet industry was well-established, accounting for 8% of the world's sugar production.

132   *The sugar cane industry*

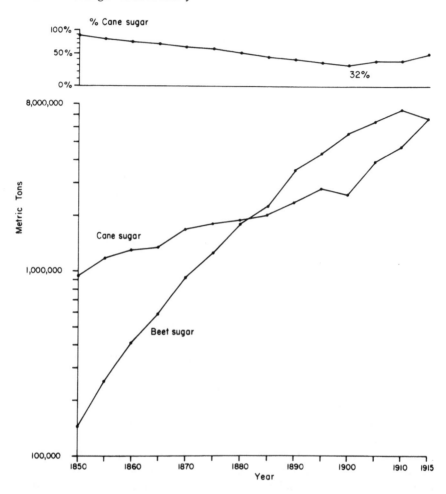

Fig. 6.1 Cane and beet sugar production, 1850–1915

Note: the data on cane sugar are a very good approximation of exports. They do not include, and I have not tried to estimate, the non-centrifugal production of India, China and of Latin America that did not enter the international trade. Accurate data on the production of all types of cane sugar in the nineteenth century are certainly not available, and therefore in this comparison with beet sugar production cane sugar is underrepresented.
Sources: beet sugar: data from Deerr 1949–50: vol. 2, 490–1. The data record production. Cane sugar: from data given in this volume, increased by 10% to account for unlisted minor exporters.

Tariffs and taxation were to be major factors in the competition between the two forms of sugar throughout the remainder of the century. The British market, the largest in the world at this time and hence a major player in the international sugar trade, became increasingly open after 1845 as Britain reduced sugar duties in the interests of free trade. Between 1874 and 1901, Britain did not levy tariffs or duties on either cane or beet sugar. In the meantime, several states on the continent of Europe were subsidizing the export of beet sugar. The major beneficiaries of this state of affairs were the British consumers who bought their sugar very cheaply. The Secretary of State for the Colonies, Lord Derby, apparently gave little thought to the welfare of the colonial populations under his charge, when, reviewing the sugar trade in 1876, he commented that he saw little reason to complain if France, Germany and Austria chose to tax their citizens to give the British cheap sugar (Beachey 1957: 55). This hands-off attitude had the result that by the end of the century, two-thirds of the sugar entering Britain was beet sugar.

The rationale for the promotion of the suger beet industry lay in the depressed state of European agriculture. The wheat exports of Russia were joined in western Europe by grain from the Americas and prices became even further depressed. Governments turned to beet as an alternative cash crop. They levied taxes on the raw materials and equipment of the beet industry, rather than on the finished product, and payed bounties on the export of beet sugar. The method of taxation had the effect of encouraging efficiencies in manufacturing, and in Germany the rate of extraction of sugar from beets improved from 5.8% in 1836 to 15.6% in 1906 (Timoshenko and Swerling 1957: 236). In Germany, bounties on exports were related to the sugar yield of beet, and by the 1880s the improvements in yields led to a significant increase in the bounty for exporters (Perkins 1984: 32). Germany emerged as Europe's major producer and the effect on sugar prices can be seen in Figure A.2.

The popularity of cheap sugar in Britain, Britain's continuing adherence to the doctrine of free trade and the persistent depression in European agriculture combined to defeat the first attempts to rationalize a system of trade that was flooding the international market with sugar at prices well below the cost of production. Only at the very end of the century did the political and economic situations change. In 1895, Britain acquired in Mr. Joseph Chamberlain a Secretary of State for the Colonies who had at one time invested in the West Indies and was familiar with West Indian problems. He became convinced that the continental bounties were the cause of the economic problems of the British colonies and that some defense of their interests was in order. There were problems at home, too, because the refined continental sugar could compete successfully with Britain's own refiners. Chamberlain had to overcome the opposition of the free trade Treasury as well as the supporters of the public's right to cheap sugar, but in 1895 he succeeded in persuading the government to establish a royal commission with a mandate to inquire into the effect of the foreign competition on the West Indian producers and to recommend remedies. The commission worked quickly, and its report, delivered in 1896, not only confirmed Chamberlain's position but the commissioners were unanimous in urging the government to convince the continental countries to abandon the bounties (Beachey 1957: 150–5). Meanwhile, opinion on the continent, too, was changing. The cost of the bounties had escalated into a heavy charge on national accounts, while agricultural prospects had improved, in part because of the introduction of tariff protection. The United States and British India had already introduced tariffs against subsidized sugar. The time was ripe for negotiations between cane and beet producers.

The Brussels Convention of 1902 abolished the sugar bounties, and so ended the unequal competition between cane and beet sugar. It did not end the politicking in Britain between free traders and factory owners who wanted cheap sugar and imperialists who supported the West Indian producers; nor did it end the expansion of the sugar beet industry. The United States and Russia, for instance, increased their acreages to supply their own growing domestic markets and the beet industry spread to Scandinavia and southern Europe also to supply domestic needs. But the Convention arrested the decline of cane sugar. Whereas in the years immediately before the Convention beet sugar contributed 52% of the world's sugar production, by 1914 this share had dropped to 48%. (Beachey 1957: 137–74; Chalmin 1983: 14–26; Chalmin 1984: 9–19; Timoshenko and Swerling 1957: 235–40).

Technology

By the year 1800, the sugar industry in the western world was nearly 1,000 years old, but during this vast span of time it had been extremely conservative in its adoption of innovations in the milling and manufacture of sugar. Undoubtedly, the most important innovation in milling had been the three-roller mill. In the boiling and curing houses no other single innovation stands out in quite the same way but among the most significant innovations surely were the battery of cauldrons, the Jamaica train and the discovery at some uncertain date of the role of damp clay in whitening sugar. After 1800, however, technological innovation becomes a major theme in the history of the industry. The changes in milling and in the methods of manufacturing sugar reflect, of course, the extraordinary surge in knowledge of engineering and chemistry that characterized the industrial revolution. Innovations in one sphere of activity were adapted to another, and the processing of cane sugar borrowed and benefited from the experiments in the sugar beet industry. Technology became a determining agent of change: new technology created larger mills and factories than the industry had known before and they demanded accordingly a more extensive hinterland of cane-pieces. This led to an increase in the optimum size of plantations and so impelled a consolidation of landholdings that in turn had repercussions on settlement patterns. The cost of the technology was also at a new level of magnitude, and problems of finance helped move ownership of the industry from planter families to corporations.

Steam power was the first innovation of the nineteenth century. In fact, planters had begun to investigate the use of steam power to turn their mills as early as 1768 when the Jamaican Assembly recorded "It has appeared to the Committee that John Steward has completed a mill for grinding sugar canes by applying the force arising from the steam of boiling water instead of cattle," but the experiment does not seem to have been successful (Deerr 1949–50: vol. 2, 549–51). It was not until 1797, in Cuba, that steam power was used to mill cane, but the mill did not function for more than a few weeks (Fraginals 1976: 36). Deerr (1949–50: vol. 2, 552–3) has been able to show from an examination of the order books of two English engineering firms that by 1820 steam had become an accepted means of powering sugar mills: one of the companies exported 114 mill engines between 1802 and 1820, the other 80 between 1813 and 1817 to customers in the West Indies and Latin America. Despite the advantages steam promised, the rate of adoption of steam was very slow until at least mid-century. In Cuba, by 1846 only 20% of the mills used steam, but by 1860 70% did (Marrero y Artiles 1972–84: vol. 10, 159). Barbados had only one steam mill in 1846 (Deerr 1949–50: vol. 2, 553) and of its 500 estates in 1862, only 30 used steam (Beachey 1957: 63). In Pernambuco steam mills were the exception even long after 1850 (Galloway 1968: 296). Capital, of course, was difficult to raise at a time of declining sugar prices and when planters were confronting the problems of abolition but there were also good reasons for this caution before making the investment in steam. If well-maintained, steam mills were reliable but good servicing by qualified mechanics could not be assured and spare parts might take weeks to arrive. A mill out of order in the harvest season was a problem no planter wished to confront. On some islands, according to Beachey (1957: 64) the shortage of fresh water was a problem. Another basic difficulty lay in connecting steam engines to three-roller mills: they were not a harmonious fit and it was common for powerful steam engines to damage mills. The answer lay in a new design of mill.

The sugar mill began to assume its modern form in 1794 when John Collinge, an English axle-tree maker, produced a design that consisted of three hollow iron rollers, laid horizontally, held in place by an iron headstock with the upper roller pressing down on the two lower ones. Viewed from the side, the setting of the rollers in the headstock marked the angles of an isoceles triangle. Mills of this design soon began to be exported from England, and Deerr (1949–50: vol. 2, 537) records that one company during the years 1813–17 exported sixty-three of them compared to only eleven of the vertical three-roller type. Throughout the nineteenth century the design was modified to make it more efficient. Deerr (1949–50: vol. 2, 537–49) provides a detailed review of these modifications that consisted in the main of improvements in engineering of the headstock. The new design had two important advantages: (1) as heavy pieces of machinery they could stand the power that steam provided, and (2) they permitted multiple milling of bagasse. Millers had long appreciated that multiple milling would increase the extraction of juice but with the old three-roller mills the labor costs made such an exercise uneconomic; now with the new design the cane could be made to flow through a succession of linked horizontal three-roller units. These mill trains or mill tandems could consist of three to seven sets of three-roller units. Further refinements were added to milling. In some mills, the cane was made to pass between two heavily grooved rollers or crushers that extracted between 40% and 70% of the juice before going through the mill tandem where the remaining juice was extracted. In other mills the cane first passed under knives mounted on a cylinder rotating at high speed and then through a shredder which between them broke down the hard structure of the cane, rupturing the cells, and emitting the cane as a course dense fibrous blanket which was fed down a chute into the mill tandem. This machinery took several decades to evolve. A patent for revolving knives was issued in 1854, although this piece of equipment did not become standard until the 1920s in Hawaii and later elsewhere. The earliest patent for a form of shredder dates to 1850, but the first successful version was installed in Louisiana in the late 1880s. The grooved crushers also appeared in the 1880s (Deerr 1949–50: vol. 2, 546; Meade and Chen 1977: 53–60). The mill tandems did not come into widespread use until the early years of the twentieth century, first in Hawaii and later elsewhere. Since the early twentieth century, there have been no radical innovations in the milling of cane: the arrangement of three-roller units in tandems persists to the present day (Meade and Chen 1977: 60).

The work of the boiling and curing houses also underwent a revolution that led to the replacement of the standard eighteenth-century system described so carefully by those two observant clerics, Fathers Antonil and Labat. The inspiration for the new processes and new equipment came from refineries in Europe and research in the sugar beet industry. The two inventions of fundamental importance were the vacuum pan and the centrifugal.

The vacuum pan exploits the well-known relationship between the boiling point of liquids and atmospheric pressure. Under artificial conditions of reduced pressure, it lowered the boiling point of the cane juice and thus greatly economized on fuel. It was to be used both on the plantations in the primary conversion of cane juice into sugar as well as in the refineries. The vacuum pan dates from 1813 when Edward Charles Howard, a brother of the 12th Duke of Norfolk, patented his invention, but Howard's sound predictions of its usefulness to the sugar industry took time to prove correct. By 1827, of the 100 refiners in and around London, only 10 had installed these pans. A refinery in Vienna, laid out by Howard in 1818, was the first on the continent of Europe to use the vacuum pan, and in Magdeburg in 1835 it appears for the first time in the sugar beet industry. Sir John Gladstone, father of the prime minister, became the first sugar cane planter to use a vacuum pan when in 1832 he installed one on his estate, Vreed-en-Hoop, in Demerara (Guiana). The various enlargements and improvements made to Howard's original design during the course of the century led to the development of multiple-effect evaporation in which two, three or more pans were linked in a series so that vapor from one pan could be used to evaporate liquid in the next. Norbert Rillieux built the first multiple-effect vacuum evaporator in Louisiana, in 1845. Howard's and Rillieux's inventions established the principles of vacuum boiling that apply today in the sugar as well as other industries (Deerr 1949–50: vol. 2, 559–73; Meade and Chen 1977: 46). Despite the advantages of the vacuum pan, it diffused slowly through tropical America in part because of its cost – in the 1870s the installation of even a small apparatus required an investment of between £40,000 and £50,000 (Beachey 1957: 68) – and in part because it was suited to a larger scale of enterprise than most of the existing plantations. Its widespread adoption came with the reorganization of the industry around central factories or *usinas*.

Until the invention of centrifugals, cane planters depended on the age-old method of natural drainage to separate the molasses from the crystals. Centrifugal force was first used in the sugar beet industry during the 1840s to help extract juice from the beet pulp and was borrowed by the sugar cane industry. In centrifugals, the massecuites (the syrupy mass from the vacuum pans) is placed in "baskets" that can be rotated at high speed. Screens within the "baskets" retain the crystals but allow the molasses or "mother liquor" to escape. During the first half of the century, the sugar beet refiners began to use various machines that relied on suction to draw off the mother liquor and for a time at least preferred these machines to centrifugals, but they never found favor on the cane plantations. Those planters who could afford centrifugals quickly installed them because they could purge the massecuites at the rate of 2,300 to 2,700 kg. per hour compared to the two to three weeks' drainage required, and the centrifugal yielded a drier sugar (Deerr 1949–50: vol. 2, 548, 573; Meade and Chen 1977: 45, 335).

There is some discussion over who invented the centrifugal. Deerr (1949–50: vol. 2, 573–5) has reviewed the evidence of the patents. In 1837, Penzoldt took out a patent in France for a machine generating centrifugal force, but to dry textiles. It was modified, adapted to the needs of the sugar industry, and by the 1850s, centrifugals derived from Penzoldt's patent were in use in India and they also became popular in Germany. Despite this success, in 1843, two further patents were issued for centrifugals to purge sugar massecuites, one in England, the other in France. The French patent, Seyrig's, did describe a practical machine that French engineering companies sold in large numbers to planters in French West Indies, in Mauritius and Réunion, but Deerr records nothing more about the English patent. The English, in fact, were not successful in this branch of engineering, for a second English patent for a centrifugal, taken out in 1850 by none other than Henry Bessemer who later became famous for his contributions to the iron and steel industry, did not enter production even though the prototype displayed at the Great Exhibition of 1851 attracted a good deal of attention. One of Bessemer's innovations – the suspended centrifugal – appears in later designs of centrifugals, particularly in that of David McCoy Weston who during the early 1850s was experimenting with a centrifugal at a sugar factory in Hawaii. In 1852, a Mr. Johnson, otherwise unknown in sugar mill technology and whom Deerr surmises may have been an agent for Weston, filed in Hawaii a patent which he later withdrew for a suspended machine. On the expiry of Bessemer's patent in 1867, Weston took out a patent in his own name that included suspension in the design, and it went immediately into production. Meade and Chen (1977: 46–7, 334) in contrast to Deerr are direct in crediting Weston with the patent for a suspended centrifugal in 1852 and with applying the design to practical work in Hawaiian mills in 1867. According to them the centrifugals that came into general use were known well into the twentieth century as Weston centrifugals and little has changed in the fundamentals of the design down to the present.

In addition to the vacuum pan and centrifugal, other aspects of nineteenth-century science and technology had the effect of increasing the optimum size of plantations, and helped to make the traditional mills and purging houses obsolete. With the rise of industrial chemistry, the quality of the chemical control in the manufacture of sugar greatly improved, at first in the beet industry and then after about 1870 in the cane industry. Only a large enterprise could afford to maintain a resident chemist, as indeed only such an enterprise could afford an engineering staff to keep the ever more sophisticated machinery in first class order. Railways revolutionized transportation on the plantations: steam locomotives running on narrow-gauge railways not only displaced the ox- and mule-carts but greatly enlarged the range over which a mill could draw for its supplies of cane.

As the innovations in the industry accumulated, the case for a rationalization of sugar production through a concentration of landholdings and the replacement of small mills by central factories became stronger, and increasingly planters were forced to face the fact that they would no longer be able to combine the roles of landowner and industrialist, chemist and engineer, agronomist and trader, and still hope to sell high quality sugar on the international market.

A central factory could effectively exploit the new technology, fund the employment of specialized personnel and achieve economies of scale. The idea of a central factory for the cane industry was first mooted in France in 1840–1 and the French began to put it into practice in their West Indian colonies in 1844–5 (Schnakenbourg 1980: 200–2; 1984: 88). The task of reorganization was, however, enormous: in the 1830s and 1840s, there were about 6,500 sugar estates in the Caribbean, with 1,500 each on Cuba and Puerto Rico, with many hundreds more in Central and South America. In some regions, such as Pernambuco (Galloway 1968: 295), sugar planters continued to install old-style mills and purging houses long after knowledge of the new equipment and the advantages of the central factory system had become widely known. Understandably, the switch to the central factory system was very gradual and in a sense is still not complete today as old-style mills continue to operate in the backwoods and sell their non-centrifugal coarse sugar – the *rapadura* of Brazil and *panelas* of Spanish America – in small country markets.

Economic as well as social factors delayed the adoption of the new system. The central factory system did call for large amounts of capital, usually more than individual planters could raise, and the sugar cane industry was not a terribly attractive investment to outside entrepreneurs at a time of free trade, competition from beet and declining sugar prices. Planters unwilling or without the resources to face the trauma of change could continue to supply the internal, local markets with non-centrifugal sugar. Initially, some entrepreneurs in an effort to reduce the start-up costs of central factories hoped to limit their investment to the factory and its machinery, contracting with surrounding landowners for supplies of cane rather than buying land. But such arrangements were difficult to work as planters, with their old-style mills in working order, were in a position to blackmail factory owners into paying very high prices for the cane, and the factory owners found that to survive they had to own land as well. There were also many examples of local initiative of the sort in which groups of neighboring planters jointly raised the capital for a central factory that would serve them all. The social impediment to the central factory system involved the class status of the sugar planters. Owning their own land and mills, the planters had formed the elite of society and saw the central factory owners forming a new and much more restricted elite. In some regions,

Table 6.2 *The efficiency of sugar factories*

	Primitive plant	Modern factory
Tons cane crushed per hour	1.25	227.00
Mill extraction (%)	65.00	95.30
Tons cane per ton commercial sugar	12.00	8.30

Date of comparison: 1917.
Source: Barnes 1974: 8.

as matters turned out, the situation became even more humiliating for the planters than they had feared: because of the very high costs of building a factory, the capital came from abroad so that the planters found themselves in the position of either entering into contracts to sell their cane to foreign-owned companies or selling their land and retiring from the sugar scene.

The planter class fell victim to the industrial revolution and technological progress, and the landscape its style of sugar production had created throughout such large reaches of tropical America was to be profoundly changed to meet the demands of the new order. The plantations were not only units of production, organizing the land use and flow of traffic from field to mill, but, as the planters had also decided where the slaves, and later the free workers, were to live, they were also functional units in the settlement pattern. The introduction of the central factory system meant that there would be fewer decision makers organizing land use and they would control a greater extent of territory. It was difficult to find another use for the redundant mills and purging houses which soon began to crumble into ruins; often also the plantation houses were abandoned as the planters sold their land and moved away. There was little reason for the rural workers to remain where the old system had placed them. Their modest homes did not represent very much in the way of fixed assets, and, indeed, the wooden cabins that were a common form of housing in the West Indies could be quite easily transported to another site. They began to gravitate into larger nuclei, around the central factories, in settlements that were to amount in some cases to company towns. A nostalgia arose among descendants of the old planter class of the cane-growing regions for the way of life that was passing as one arose in the South of the United States for antebellum times. No one has expressed this nostalgia more strongly than Gilberto Freyre (1933) whose lengthy defense of the contribution the sugar planters of northeast Brazil made to the evolution of society has now become a classic work in Brazilian historiography. It is widely known in the English-speaking world as *The masters and the slaves*.

By 1914, the transition to the central factory system was well underway: the numbers of old-style mills were declining, and central factories were no longer novelties in most cane-growing regions. These improvements in the milling and manufacture of cane sugar, and the economies of scale they permitted, were an important factor in enabling cane sugar to compete with beet (Fig. 6.1).

Cane breeding

The breeding of new and better varieties of sugar cane was another means by which the industry met the competition from beet. Until the late nineteenth century, the industry substituted one variety of naturally occurring cane by another when the necessity and/or opportunity arose, just as Otaheiti or Bourbon cane replaced Creole in the years around 1800. Otaheiti cane gave good yields but it was found to have an important weakness: it was susceptible to a variety of diseases when grown under intensive conditions, as on plantations. In the 1840s, disease struck the cane fields of Mauritius so severely that the planters decided to replace the Otaheiti cane, and their search led them to the "noble" varieties of the Cheribon group of Java. As Otaheiti cane "failed," to use the jargon of the industry, in various places around the world – Brazil in the 1860s, Puerto Rico after 1872 and in the other Caribbean islands after about 1890 – Cheribon cane took its place. There was no guarantee, of course, that Cheribon cane would prove any more disease-resistant than Otaheiti cane, and so Mauritius continued to sponsor the search for other varieties long after Cheribon was in commercial cultivation. The specimens of sugar cane from South China and the various islands of Southeast Asia and the Pacific that arrived in Mauritius were propagated at the Pamplemousse Botanical Gardens and had by the 1880s made the cane collection there possibly the largest in the world. Likewise, the Queensland Department of Agriculture established a collection of Melanesian varieties of sugar cane at its Mackay State Nursery. The diffusion of the different varieties of cane about the globe was in itself a hazard to the sugar cane industry because collectors did not screen their samples for disease. The fungus that caused "red rot" or "rind disease" reached the British West Indies in 1882 in a case of samples from Mauritius, and soon thereafter attacked the Otaheiti cane on the islands. Mosaic disease, of Melanesian origin, came to Argentina in cuttings sent from Java.

The vulnerability of sugar cane to disease meant that the policy of substituting a failing variety with a new, naturally occurring one could give the industry no security, and that a better approach to the problem might lie in the scientific selection and breeding of new varieties of cane. Such research was rather slow in starting. It was the conventional wisdom of those familiar with sugar cane that it was sterile and did not set seed. The Creole and Otaheiti canes – the only varieties known in the Americas until the second half of the nineteenth century – are indeed male-sterile, but other varieties do set seed, and the discovery of the fertility of sugar cane depended on familiarity with these varieties as well, it seems, as compelling economic motive.

Seedlings of cane were first identified in May 1858 in Barbados, but the discovery did not lead to any productive experimentation and appears to have been forgotten or at least was ignored as the Black Cheribon variety was performing extremely well on the plantations. However, the outbreak of Sereh disease in Java renewed interest in disease-resistant varieties. The rediscovery that cane was fertile took place almost simultaneously in Barbados and Java in 1888, and the scientific breeding of sugar cane began immediately thereafter. There were six breeding stations in existence by 1900, four more by 1914, and soon thereafter virtually every cane-growing region of any significance established a research station. The beneficial results for the sugar cane industry were soon to appear, and have continued to flow from this line of research ever since. It has been possible to develop varieties that are resistant to diseases, that suit different soil, water and climatic conditions, that germinate quickly and so reduce the weed cover and the labor costs of weeding, that give more and better ratoon crops and that are easier to mill. After the First World War, improved cane stock from the research stations rapidly replaced the natural varieties in commercial sugar cane production.

This varietal diversity and the frequency with which new improved varieties were, and are, introduced, along with the central factory system, helps distinguish the modern sugar cane industry from the old. Plant breeding and technological innovations, taken together, in large measure explain how the sugar cane industry was able to take advantage of the small measure of encouragement the consuming countries such as Great Britain began to give it in the late nineteenth century and so to compete successfully with beet (Timoshenko and Swerling 1957: 125–32; Stevenson 1965: 39–71). The First World War had important consequences for both beet and cane sugar. Beet production declined sharply, not least because of warfare in the beet-growing countries. Britain in 1914, finding itself at war with Germany and Austria–Hungary, its major suppliers of sugar, determined to revive suppliers of sugar within its empire. By 1920, the sugar cane industry had recovered to the extent that it accounted for 70% of the world's sugar production.

7
The geographical responses to the forces of change: 1790–1914

The problem that faced the American sugar cane industry during the nineteenth century can be briefly restated: the planters had to replace slavery with new systems of labor as well as finance revolutionary innovations in milling and manufacturing at a time of declining prices caused by competition from beet sugar. The planters did have forewarning that slavery would not last indefinitely, but the full implications of the industrial revolution for the technology of the industry and the seriousness of the challenge from beet growers only gradually became apparent with the passage of time. Whatever the opportunity the planters had to lay strategy for a difficult future, the challenge of modernization was indeed a formidable one; and the ways in which they responded changed the land use, settlement patterns and ethnic composition of the population of the cane-growing regions, affecting profoundly in the process the geography of tropical America.

The key variables that determined the pattern of response in each region were the familiar ones of resources, capital, markets and government policies. In those small islands where the plantation economy monopolized the land, the planters did not experience a labor problem after abolition because the former slaves had no alternative but to work in the sugar industry; however, where there was easily accessible land, the former slaves did have the option of leaving the plantations to become small-holders. Fertile land gave higher returns per laborer employed than poor land, and hence those planters fortunate enough to be on good quality land were in a better position to attract and retain a labor force than those who worked worn-out lands. A combination of worn-out plantations close to unoccupied land had a devastating effect on sugar production as the example of Jamaica, discussed below, will show. Capital for financing the new technology flowed to where it could expect a good return, that is to regions rich in natural resources and with access to large markets. Cuba, close to the huge new North American market for sugar, was an outstanding example of this. The manner in which the various governments brought slavery to an end and their policies on post-emancipation forms of labor had an important bearing on the prosperity of the industry. British acceptance of the doctrine of free trade at a time when European governments were subsidizing the cultivation of sugar beet was very harmful to the interests of its sugar cane colonies, yet, perhaps paradoxically, the same government helped the industry by the support it gave to Indian indentured labor. In South America, the newly independent countries fostered their own industries through subsidies and protective tariffs for domestic political reasons.

The experience of each cane-growing region was, of course, unique, but underlying the detail in the record of the individual regions it is possible to recognize four distinct patterns of response to the forces of change. These can be categorized as follows: (1) the decline of the industry, (2) a gradual increase in production in a slowly changing social and organizational environment, (3) rapid expansion, financed by foreign capital and supported by demand in international markets, and (4) expansion supported by demand in national markets. The first two responses were associated with long-established sugar colonies; the third and fourth mainly with new centers of production. Where planters were least able to make the transition from the old style of sugar production to the new, the industry declined; other types of land use took the place of cane fields, and the former slaves dispersed from the barracoons to found villages of their own. Haiti, the successor state to St. Domingue, the most important of the late eighteenth-century sugar colonies, provides an extreme example of the first response. In Jamaica, too, the industry underwent a very steep decline although not to the point of withdrawing completely from the international market. The experience of Barbados illustrates the second response: Barbados was also an "old" sugar colony, but in contrast to Haiti and Jamaica its planters managed to increase production through agricultural improvements and technological innovation while making only minimal changes in the organization of the industry and without recourse to new supplies of labor. The closest parallels to the Barbados response were St. Kitts, Nevis and Antigua, all small islands and, like Barbados, subject to British policies towards labor and sugar duties. Although differing in some ways from this response, the "old" sugar colonies of Martinique, Guadeloupe and even northeast Brazil have more in common with it than with any of the other three responses. The industries of Cuba, Puerto Rico, Trinidad, British Guiana, the Dominican Republic and coastal Peru are examples of the third response: the planters were able not only to find new sources of labor but to finance the technical improvements and a great expansion in the acreage of sugar cane. The new industries of Louisiana, Tucumán, the Cauca valley of Colombia and São Paulo together with one long-established industry – that of Morelos in Mexico – fall into the fourth response of protected industries producing for national markets.

The old sugar colonies

St. Domingue and Jamaica

St. Domingue was the only sugar colony not to manage a peaceful transition from slavery to free labor, and one of the many consequences was the total destruction of the sugar industry. Social and political tensions in this class-conscious and racist society had already reached a point of great strain by 1789 when the political events in France gave the various factions a new rhetoric with which to argue their respective causes. The rich planters began to realize that their interests lay with the *ancien régime*, the poor whites wanted political power; and neither group was willing to accept the coloreds as equals in alliance against the blacks. The coloreds' references to the Declaration of Rights of Man and Citizen in support of their own economic and political aims rang rather hollow when they did not wish to extend these Rights to the slaves whom they feared. The alliances, the tolerated lines of authority, the tacit assumptions and the civilities that hold a society together began, in St. Domingue, to come asunder. The slaves after observing two years of turmoil and political debate, finally took advantage of the power vacuum to revolt on the night of August 22, 1791.

The revolt could not be contained. Both the British and Spanish, alarmed by the precedent, sent troops but their interventions were ineffective, and the army ordered by Napoleon to St. Domingue failed to impose order, defeated by disease and regiments of ex-slaves. The French finally withdrew on November 30, 1803, in twenty ships with over 18,000 refugees on board, leaving the leaders of the former slaves to proclaim the independent republic of Haiti on January 1, 1804.

The warfare had ruined the economy. For the next twenty years the rulers sought to restore to operation at least some of the sugar plantations, but the labor problem proved to be insuperable. There was a great deal of vacant land in Haiti, both the abandoned cane-pieces in the plains as well as extensive ranges of uncultivated hills, and the ex-slaves who generally preferred to be their own masters rather than to work for others moved onto this land to eke out a simple, subsistence existence. The danger of empressement into one or other of the marauding armies of the contenders for power and the risk of malaria, which seems to have increased as the irrigation works fell into disrepair, encouraged a migration from the plains to the hills. Henri Christophe, ruler of the north of Haiti (1807–20), had some success with enforcing his doctrine that agricultural workers should remain on the estates of their birth, and briefly thereby increased sugar exports. In the south of the country, President Pétion (1807–18) followed a policy of distributing land in small plots to ex-slaves. Those landowners who tried to maintain at least some commercial production through sharecropping arrangements found that the sharecroppers behaved as if the plots of land were their own to grow as little or as much of whatever crops they wished. By the mid-1820s, the sugar industry had ceased to exist and commercial agriculture in Haiti was reduced to the small crops of cotton, cocoa and coffee that peasant cultivators could grow (Leyburn 1966: 32–79; Ott 1973: 188–200; Moya Pons 1972: 193–7).

There was no recovery from this situation during the remainder of the nineteenth century as successive governments were unable to provide the basic logistical support, whether transport, a respected currency or even rule of law, that large-scale commercial agriculture required. The transformation of the human geography of the former sugar colony between 1789 and the 1820s could hardly have been more complete: the leading sugar producer of the Caribbean ceased to export sugar, large estates gave way to small-holdings, a slave population became a peasantry which in large part abandoned the plains in favor of the hills. Haiti imported little and exported little, and did not participate in the major commercial trends of the nineteenth century.

The geographical circumstances of Jamaica were very similar to those of Haiti in that the sugar industry only occupied a part of the land area of the colony before abolition, and therefore former slaves could abandon the plantations in favor of a peasant existence. Jamaica, however, did not experience a violent revolution, and the colonial government was able to guarantee law and order, and provide a financial and administrative infrastructure that was modern by the standards of the time. Even so, the Jamaican sugar industry underwent a major decline during the nineteenth century with exports falling from a peak of about 90,000 tons annually between 1805 and 1807 to about 25,000 at mid-century and 15,000 at the end. This decline was also reflected in the number of plantations: 670 at abolition, 202 in 1880.

Emancipation forced a change in the method of paying for plantation labor and inaugurated a labor market in Jamaica. Where before planters were able to buy slave labor on credit through their London agents, settling the debt later in sugar, now they had to have cash on hand – in advance of the harvest – not only to offer cash wages, but wages at a sufficiently high level as to induce the former slaves to continue to work on the plantations rather than become peasant cultivators. Jamaican planters were notoriously short of cash, could not afford to pay well and, what is more, were often in arrears. The planters were in fact unable to offer an attractive deal to their former slaves who quickly abandoned the plantations in very large numbers. Nor were they able to meet the costs of imported indentured labor. The costs were high, including the sea passage, wages and expense of administration, and, moreover, indentured workers had to be supported the year round, in good seasons and in bad when local labor could be dismissed. Indentured workers were not a cheap alternative to paying good wages.

The shortage of labor drew two responses from the planters: they decreased the acreage of cane and endeavored to increase the productivity of the workers through agricultural and technological improvements. Planters replaced hoes with plows, gave more emphasis to ratooning because ratoon crops required less work than planting new cane and managed after the 1840s to reduce the labor costs of planting and caring for an acre of cane by about 65%. They raised yields by paying increased attention to manuring, applying local and Peruvian guano when their own livestock could not provide sufficient quantities. There was even some discussion of the merits of digging in the bagasse along with the animal manure and using imported coal in the furnaces. The Otaheiti and Cheribon canes increased yields. By mid-century about one third of the mills were powered by steam while nearly half of the mills had been equipped with the new horizontal rollers (Hall 1959: 71). Skillful management as well as capital resources were necessary to make these adjustments, and not all planters were able to meet the test: between 1836 and 1846 as many as 157 sugar estates were abandoned (Eisner 1961: 198).

The problem of raising capital for the plantations became particularly acute. In a reversal of eighteenth-century experience, absentee ownership was now an advantage – provided, of course, the absentee employed a good manager to administer his properties – because the absentee in Britain had easier access to financial institutions than a resident owner. For years, however, many of the plantations had been unable to repay in full the advances from their agents and many also had in more prosperous years been charged with legacies which the present revenues could not support. Owners of estates encumbered with debt were in no position to raise capital for improvements. Indeed, the debts and encumbrances often exceeded the probable market value of the land and buildings; creditors could expect to get possession only after lengthy legal wrangles and there was little likelihood of buyers coming forward for plantations in this condition. There was, in fact, a good prospect that many of these debt-encumbered plantations would soon also be abandoned by their owners to the further detriment of the sugar industry and local economy.

In order to break through the legal impediments in the way of transferring the ownership of plantations, in 1854 the British government passed the West Indies Encumbered Estates Act which enabled either owners or creditors to apply for estates to be sold by judicial decree. London left to the individual colonies the decision whether or not to adopt the Act. Jamaica

The sugar cane industry

adopted it in 1861, and 148 estates were sold before the Encumbered Estates Court ended its work in 1892. The new owners received the estates free of encumbrances and with clear title, and thus were in a position to carry out improvements. The Act also made for the concentration of landownership: in Jamaica, one petitioner bought twelve estates, another nine. Several British companies bought estates (Beachey 1957: 36).

The Encumbered Estates Act removed one difficulty in the way of the improvement of the Jamaican sugar industry, but it did not lead to the modernization of the industry through the introduction of the central factory system. Other difficulties still stood in the way. The plantations were small – the average size in 1880 was 185 acres – and the individual planters were loath to lose their independence by becoming cane-farmers or by merging into an association to found a central factory. The introduction of the centrifugal, which separated more efficiently and thoroughly the molasses from the muscovado sugar than simple drainage, gave a fillip to the profitability of the old purging houses: well-drained molasses was important to the flourishing Jamaican rum industry and, moreover, there continued to be a market for muscovado in the United States. In 1872, the change in American duties on sugar to favor domestic refiners resulted in lower charges on imported unrefined sugar, a change from which Jamaican and other British West Indian muscovado producers benefited. By the end of the century the U.S.A. was taking over two-thirds of the sugar production of the British West Indies (Beachey 1957: 128–9). The low price of sugar, especially after 1884, and the uncertain prospects for cane sugar in the British market before the Brussels Convention of 1902 made central factories too risky an investment in Jamaica.

Centralization began in earnest in the first years of the twentieth century. In 1902, the Jamaican government passed a law that encouraged the building of central factories, and neighboring estates began to cooperate and re-equip. The British demand for cane sugar during the First World War greatly speeded the process and in 1930, at the onset of the Depression, there were forty central factories serving thirty-nine plantations with an average size of 661 acres. It is, perhaps, a reflection of the poor competitive position of the Jamaican industry that they were owned by Jamaicans with the exception of three that belonged to the United Fruit Company. But during this lengthy transformation from slave-worked, family-owned plantation to modern central factory, sugar's contribution to the total value of Jamaica's exports sank from 60% to 12% (Eisner 1961: 202–3, 206–9, 238).

As the sugar industry declined, so other forms of land use and land tenure grew in importance and completed the reformation of the economic geography of the island. The major development was the emergence of a peasantry which began on a large scale immediately after the end of apprenticeship in 1838. Some ex-slaves rented land from planters, a few even had the resources to buy plots for themselves, and yet others found homes in the so-called free villages founded on land the Baptist missionaries had bought. The extensive crown lands as well as the abandoned plantations provided plenty of scope for squatting. The new peasants at first cultivated food crops which they sold either to the estates or to the towns. With the decline of the estates, this market shrank, and peasants turned of necessity to export crops. Much of the increase in coffee exports in the mid-years of the century was due to peasant production. They also grew arrowroot, fustic, coconuts, sugar and ginger, and collected logwood, beeswax and honey. The peasant economy depended on access to markets, and in those parts of the island where transportation was poor, many continued to work for part of the week on the sugar plantations.

Statistics on the size of peasant plots are hard to come by for these early years, but by 1860 there appear to have been about 50,000 holdings of under 50 acres. After 1860, the peasant class grew rapidly, and by 1900, there were just over 100,000 holdings of less than 5 acres. A key factor in the expansion was the insurrection of 1865, the result of the hardship caused by drought, higher prices for imported goods – in part at least a consequence of the Civil War in the United States – and maladministration. Britain responded by abolishing the planter-dominated Assembly and installing Crown Colony government. The new governors acknowledged the important role the peasants had come to play in the economy, and began to encourage their activities through new financial institutions designed to assist them, by agricultural education and by promoting policies to help the peasants acquire clear title to the land they cultivated. These developments in favor of the peasants were supported by the cash returns from the cultivation of bananas for the United States market. So profitable, indeed, did the banana trade become that in the 1880s and 1890s many sugar estates turned to banana cultivation. In subsequent years, there was a tendency for estates to switch their emphasis back and forth between the two crops depending on where the greater profit lay, although the fixed costs of the sugar industry favored persistence with sugar even when prices were low (Eisner 1961: 205–35).

Slow change in the old colonies

Barbados provides the outstanding example of continuity in response to the challenges of the nineteenth century: the organization of the industry, the settlement pattern and rural economy of the island remained astonishingly stable, and with the obvious exception of the new form of labor brought by the abolition of slavery, it was only in the late nineteenth century that major changes began to appear in the organization of the industry and in the landscape of the island. The planter class continued the pattern of rational decision making that had characterized its behavior in the eighteenth century, adopting innovations only when it was useful to do so. Through the introduction of the new horizontal design of mill, and of Otaheiti and later other varieties of cane, by intensive manuring, including the importing of guano, the planters were able greatly to increase the output of sugar (Fig. 7.1) even before central factories appeared on the Barbadian scene. Of crucial importance to the course of events was the size of the island, and of secondary importance were the characteristics of landownership and land tenure in the island.

In Barbados, emancipation did not create a labor problem. Rather, in the years following the end of apprenticeship in 1838, the Barbadian planters benefited from abundant and cheap labor. Barbados is small, and virtually all of its 106,000 acres are suitable for plantation agriculture. The planters monopolized this land with the exception, by 1850, of only 2,500 acres which were then in the possession of small-holders. As there was no crown land or rugged unoccupied interior onto which the former slaves could move, they had no opportunity to acquire land unless planters were willing to sell them some of their acreage in small lots, and this the planters did not do in a significant way until after the collapse of sugar prices in 1884. In other words, in the years immediately following 1838, the former slaves had no alternative but to remain on the sugar plantations, unless, that is, they could find some means to emigrate which very few managed to do (Levy 1980: 80–3).

The planters now had to pay wages to their laborers but they attempted to reduce their wage bill and at the same time secure a permanent reliable work force for their plantations by the introduction of what became known as the Tenantry System. A planter allowed former slaves the use of a house and small plot of land, at no rent, in return for the exclusive use of their labor at a stipulated wage which was 25% below that paid to non-tenantry labor. The evidence suggests that planters increased the allocation of land when converting the slave yards into tenantries, but even so the plots the tenants occupied usually covered less than one quarter of an acre although there were plots of up to half an acre on some estates. Discontentment among the tenants led to planters charging a rent for house and plot in return for an adjustment in wages. The tenants did not have security of tenure and those who failed to meet their obligations to the planters were liable to eviction (Marshall 1975: 86–7).

The ease with which the Barbadian planters were able to move from slave to free labor ensured that their production costs were among the lowest in the British West Indies. At times of marked declines in the price of sugar they were able to reduce the rate of daily pay and still command a labor force of adequate size. Improvements in the efficiency in the use of labor combined with the natural increase in the population led to

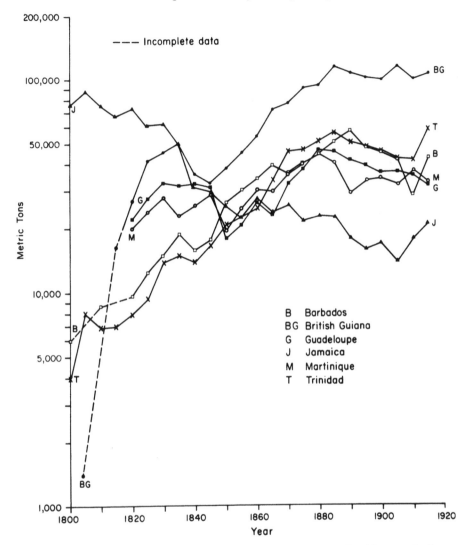

Fig. 7.1 Sugar production in the major English and French Caribbean colonies, 1800–1914 (five-year averages)

Sources: Barbados, British Guiana, Guadeloupe, Martinique and Trinidad: Deerr 1949–50: vol. 1, 193–4, 201–3, 235–6. Jamaica: Eisner 1961: 240–3.

unemployment, and rural poverty became a serious social problem. The 1871 census showed that a quarter of the total population was unemployed. Official discouragement of emigration turned in the 1860s into encouragement, with the result that between 1861 and 1903 net emigration was perhaps as high as 103,500. The outflow of labor to wherever work was available around the Caribbean and elsewhere not only eased the population pressure in Barbados but affected the island in a second way: the remittances of the emigrants raised the purchasing power of their families at home (Levy 1980: 110, 134–5; Marshall 1975: 94–5).

Despite the remittances, the records indicate only a modest increase in the number of small-holders between the 1850s and the 1880s. Both the high cost of land and the fact that planter interest still controlled access to it worked against the establishment of a peasant class. During these three decades, when plantations were sold as a single lot, the average price per acre was about £60, and plantations usually were sold as a single lot. This appears to have been a matter of policy. Barbados did not place itself under the Encumbered Estates Act, and relied instead on its own Court of Chancery to deal with insolvent estates. Of the 368 plantations sold through the Court between 1851 and 1890 not one was subdivided and offered for sale in small lots. The remittances of individual families or even groups of families were insufficient to buy estates at these prices. And when small plots of land did come onto the market, the price per acre was very high. In the 1850s, in a rare instance of the subdivision and sale of an estate in lots of 1 and 2 acres, the price asked was £70 an acre, and there is evidence to suggest that some people did in fact pay out as much as £100 and £150 per acre in their desire to become free-holders.

The great decline in the price of sugar in 1883–4 finally broke the planters' control. Unable to reduce costs sufficiently, many planters accumulated debts they eventually found they could not finance, and so their estates passed into the Court of Chancery. The Court in its turn had difficulty in selling some of the estates as single lots even with prices as low as £20 an acre: subdivision and sale or lease of small plots seemed the sensible solution. It was also politically expedient, as pressure from the descendants of the slaves for access to land strengthened with the increase in the flow of remittances, particularly after 1900. This was the so-called "Panama Money" because a good deal of it at least was earned digging the Canal, and remittances from all sources between 1901 and 1920 amounted to about £2 million. In the 1890s, four estates were subdivided, two for sale and two for lease. Between 1901 and 1920, some 4,000 acres of estate land were sold in small lots and another 860 acres was made available for lease. The small-holders gradually changed the settlement pattern by founding new or free villages as opposed to the tenantries. There were few of these villages before 1900, thirty-one more appeared on the map by 1920 and a further forty-six by 1930. It took, therefore, nearly a century before parts of the island acquired a distinctive post-emancipation landscape, but the importance of this development can be exaggerated: Barbados retained a predominantly plantation economy (Marshall 1975: 88–102; Richardson 1985).

The technological innovations in the milling and manufacture of sugar similarly had little impact in Barbados until the years around 1900. Indeed, the only innovation to be generally accepted on the island was the new design of horizontal mill which became common on the island after the 1850s (Levy 1980: 106). Throughout the century, the Barbadian planters were reluctant to transfer from wind to steam power: by 1862, less than 10% of plantations had steam mills, a proportion that increased to just over 30% by 1914. In 1897, when the West India Commission surveyed the state of sugar industry in the colony, it reported only nine vacuum pans and there were no central factories (Beachey 1957: 62, 68; Barnes 1974: 118). There were good reasons, of course, why the Barbadian planters had not been more innovative. Their supply of labor, cheap and literally on the doorstep, gave them an enormous competitive advantage, and they adjusted their production costs at times of steep declines in the price of sugar by lowering wages, by as much as 20% in the 1890s (Marshall 1975: 92). They appreciated the advantages of the new design of mill and acted accordingly, but clearly viewed skeptically the installation of steam mills which carried a substantial initial cost as well as annual fuel bills compared to the existing windmills which had very low operating costs. As long as Barbadian planters could find a market for their muscovado sugar in the United States there was little reason to invest in vacuum pans.

The characteristics of land tenure and landownership in Barbados were conservative forces that helped maintain the distinctiveness of the island's sugar industry. The production of an estate of 500 acres or more was necessary to justify investment in a vacuum pan and central factories required the yield of several thousand acres to take full benefit from the economies of scale the new system of production made possible, but in Barbados in the late 1800s the majority of plantations covered less than 300 acres. The amalgamation of plantations would have permitted more rapid technological change but the planters of Barbados were unusual in that two-thirds of them were resident owners, many with roots in the island going back several generations, and they were most reluctant either to sell or to lose their independence through mergers. Some planters did indeed own more than one estate, but rather than being contiguous the plantations were usually in different parts of the island, a pattern of tenure which was either the chance consequence of marital inheritance or the result of an acquisitions policy of spreading the risk of crop failure by owning land in the different rainfall regimes of the island. Even when land did come onto the market, its high cost made the assembly of large units a very expensive undertaking. The West India Committee recognized this problem and recommended financial assistance for groups of planters who wished to pool their resources to establish central factories. There followed a modest, gradual amalgamation. The 440 plantations in 1897, down from 500 plantations at emancipation, declined to 329 in 1912 and 305 in 1919 (Marshall 1975: 96), but this was still a large number of plantations on a small island. Nevertheless, groups of planters gradually managed to pool their resources to build central factories and hence the Barbadian sugar industry did not pass into the control of foreign companies (Beachey 1957: 68, 125–7; Hoyos, 1978: 181).

On St. Kitts, Nevis and Antigua, too, the planters owned virtually all the cultivatable land and they were unwilling to subdivide estates for sale in modest plots. Their control of the land together with access to political power gave them the upper hand in their struggle with the former slaves over the economic direction of the islands. The ex-slaves were faced with the stark choice of emigration or of remaining on the plantations. Some indeed did emigrate after the end of apprenticeship to the new colonies of Trinidad and British Guiana but the great majority remained where they were, wage earners in the same fields and mills where they had labored as slaves, living as tenants in housing on the plantations that employed them. They sought a greater degree of independence by breaking what was at first a strong tie between place of work and place of residence, striving to establish free villages on their own lands but the gains they made were very modest, particularly in Antigua and St. Kitts. So little bargaining power did the workers have that they were unable to prevent the planters lowering the wages, by 50% over the years. Cheap labor was one advantage the planters enjoyed; a major disadvantage was the small size of their plantations which greatly reduced their ability to introduce the new technology. They adopted horizontal mills and steam power because they could make efficient use of them even on small estates, but vacuum pans and centrifugals required a size of operation that simply did not exist on these islands. Denied the full benefits of the century's technological innovations, the planters concentrated on new varieties of cane and on improvements in the fields to increase the production of sugar. They further lowered production costs by reducing the number of laborers they required through the use of plows. By the end of the nineteenth century, the plantations on these three islands still produced muscovado sugar in a manner and in a landscape very little changed from pre-emancipation days (Hall 1971: 96–127).

All, however, was not quite the same, and by 1900 the scene was set for the introduction of the central factory system. There had been a gradual amalgamation of plantations, from 127 on Antigua in 1841 to 71 in 1897 (Schnakenbourg 1984: 92) and from 137 on St. Kitts in 1856 to 54 in 1901 (Richardson 1983: 100, 136). The debt burden on the plantations had been reduced through changes in ownership. The three islands had adopted the

Encumbered Estates Act, and in Antigua no fewer than seventy-three plantations were sold through the Encumbered Estates Court. When the West India Royal Commission visited Antigua it found that nearly all the island's plantations were free of debt (Beachey 1957: 36–8). The Commission encouraged the founding of central factories, and this the planters proceeded to do after 1903 when the removal of the bounties on beet sugar gave new promise to the cane industry. Two central factories were built in Antigua: the Antigua Sugar Factory that milled its first crop in 1905 and the Bendals Factory which was an old sugar works converted to the new system of production. Planters and even some small-holders contracted to sell their cane to the factories, and the original contractors became share holders when the interest-free government grant had been repaid (Barnes 1974: 117). The St. Kitts (Basseterre) Sugar Factory Ltd. began operation in 1912. It was a public company registered in London with shares owned in England as well as by resident St. Kitts' plantation owners. It, too, contracted with planters for their cane, and by 1926 had built a railway around the island to bring in the crop (Merrill 1958: 96–8; Richardson 1983: 136).

Planters in Nevis were unable to make the full transition to the central factory system. They were handicapped by poor soil so that they were less able to pay even very low cash wages than planters on Antigua and St. Kitts, and some began to lease land on a sharecropping basis. Contracts varied considerably in their details, but metairie, as sharecropping was known in Nevis, eased the problems of cash-short landowners and gave the tenants a measure of freedom. But a fundamental problem from the planters' point of view was the size of Nevis: smallness was not an asset in this case but a disadvantage because on the limited acreage of the island they could not grow sufficient cane to justify the construction of a central factory. Those of them who wished to remain in the sugar industry had to accept the expense and inconvenience of shipping their cane across the narrow channel to the factory on St. Kitts, an ultimately uneconomic arrangement, and in the twentieth century the Nevisian industry slowly came to an end, a victim of technological change.

Martinique and Guadeloupe were also "old" sugar colonies with graphs of sugar production (Fig. 7.1) that parallel those of Barbados, St. Kitts and Antigua, but behind this important similarity the course of developments in the two French islands was rather different. The crisis began later. Indeed, the years following the Napoleonic Wars were prosperous ones for the sugar planters of Martinique and Guadeloupe, a consequence in large measure of the collapse of the St. Domingue industry, and they now became the principal suppliers of sugar to France. However, the planters did have to make a concession: while they had formerly exported to France clayed sugars, fine and white enough to compete with the varieties made in the French refineries, the refiners wanted to import only the poor quality muscovado they had formerly obtained from St. Domingue and succeeded in 1814 in persuading the government to impose the necessary duties to make the planters comply. In the 1830s, however, French-grown beet sugar began to come onto the market in significant quantities. A struggle ensued between beet and cane interests in which the cane interests hoped to convince the government to suppress the beet industry; but the advantages of beet sugar to the agricultural economy gave it very strong local support and, moreover, the refiners welcomed this new source of sugar. In 1843, the government established the policy of taxing equally the two sources of sugar – a policy which in fact was to last until the end of the century. After 1843, French cane producers knew they had to deal with beet through competition rather than government favoritism, an understanding which suggests a close connection between this tax law, the modernization of the industry through the abolition of slavery in 1848 and the introduction of the central factory system (Schnakenbourg 1980: 144–58).

In view of the preceding discussions of the changes in the sugar industry in the English Caribbean islands, the process of modernization in Martinique and Guadeloupe raises two interesting questions. How was it that the planters were able to solve the labor crisis when, as in Jamaica, the ex-slaves did have the option of abandoning the plantations? And, secondly, what was the key to the early success of the central factory system in these two islands?

At the time of abolition in 1848, there was a large amount of vacant land in both Martinique and Guadeloupe. In Guadeloupe, only 25,000 of the 145,000 hectares were being cultivated and in Martinique some 30,000 hectares out of 108,000. The vacant land was mostly at higher elevations, in the hills, but nevertheless reasonably fertile. Already, before 1848, some freed slaves had moved onto this land to become peasant small-holders, and they were joined by a good many more after abolition. Yet, after only a brief decline in sugar production in the years following abolition, the increase in exports resumed (Fig. 7.1). The planters adopted two strategies to retain the labor of at least part of their former slave labor force. One was to enter into sharecropping or *metayage* arrangements. The ex-slave was allowed the use of a house, garden and a hectare or so of land on condition that he planted sugar cane. The planter claimed a half of up to two-thirds of the cane crop as his rent and paid the ex-slave for the remainder. The ex-slave also was usually allowed to graze some livestock on the plantation lands. With the produce of the garden, money for his portion of the sugar harvest and with some return from the animals, the ex-slave lived far better than he had done before. The second solution, favored by planters who needed to raise some cash, was to sell plots of land on the periphery of the plantations. These plots were often too small to support a family, and their owners supplemented their income with day labor on the estates of their former masters. The planters in fact were able to offer sufficient inducements in the way of a higher standard of living, cash income and landownership to retain the services of at least some of their ex-slaves, but even so the labor supply was insufficient for the demands of the sugar industry and the planters had to resort to the import of indentured laborers, from India in particular (Lasserre 1961: vol. 1, 392–8; Revert 1949: 260–2, 317).

The early success of the central factory system in Martinique and Guadeloupe came about, according to Schnakenbourg (1980: 200), because of the coincidence of an idea, a commercial policy and a natural catastrophe. The beet industry in which farmers sold their beets to a nearby factory provided a model for the new system, and it was indeed a manufacturer of machinery for the beet industry, the French company of Derosne et Cail, that first tried to reorganize the cane industry as a means of opening up a huge market for its equipment, suitably adapted to cane. In 1838, it convinced a large landowner on the Indian Ocean island of Réunion to build a modern factory. The results were very good but despite the publicity put out by Derosne et Cail in 1840 celebrating the success more promotion was needed. The company proceeded to guarantee assistance to purchasers of its equipment in both mounting and operating the new equipment and indeed during the ensuing years did support in many parts of the tropical world technicians "educated in the school of beet sugar factories" (Schnakenbourg 1980: 209). The central factory system also had an effective publicist in Paul Daubrée, a scientist who worked in the Guadeloupe sugar industry, whose brochure describing the industrial organization and economic advantages of the new system, and in which he called for a "1789 industriel" in the French Caribbean (Daubrée 1841: 7), attracted the attention of senior government officials. Commercial policy and government interest came together after the disastrous earthquake in Guadeloupe on February 8, 1843. The government commission that investigated the damage saw little point in spending money to repair old-style mills and purging houses, and recommended instead the building of central factories. Equipment began to arrive during the second half of 1843, and the first of the central factories began to operate in 1844 (Schnakenbourg 1980: 200–12).

The system developed in the French Caribbean in three distinct stages: (1) origins in the mid-1840s, (2) a twenty-year period of expansion, 1860s to the 1880s, between the hiatus caused by emancipation and the collapse of sugar prices in 1884, and (3) retrenchment and reorganization at the turn of the century. In Guadeloupe, Daubrée put his theories into practice and had established two central factories by 1844; a second entrepreneur, the comte de Chazelles, organized with the support of Derosne et Cail the Compagnie des Antilles which had by 1845 four central factories under construction in Guadeloupe. The owners of these factories contracted with surrounding landowners for their supplies of cane, and even at this early date brought in some of the cane by private narrow-gauge railway. There were additionally two other factories under construction on the island by associations of planters. On Martinique, two central factories were functioning by 1845, and two more by 1847. The expansion of the system during the 1860s was greatly assisted by the founding of the Société de Crédit Agricole in 1860, known after 1863 as the Crédit Foncier Colonial. The initial purpose of this bank was to assist colonial economies to recover from the problems of emancipation, a purpose that naturally led it into the business of financing improvements in the sugar industry. Financial assistance together with the demonstrated advantages of the new system had led by the end of this period of expansion to its predominance in Guadeloupe by 1883 when 80% of the island's sugar production came from central factories, a percentage that had increased to 95 by 1886; in Martinique progress was less rapid with the new and old systems dividing production about equally in 1883. Nearly 1,000 mills had ceased to function, replaced by twenty central factories in Guadeloupe and seventeen in Martinique. The reorganization and retrenchment of the last years of the century had two important aspects. The optimum size of central factories began to increase and gradually set in motion a process of rationalization whereby the larger, cost-efficient factories put smaller rivals out of business; and in Guadeloupe the owners of the factories took advantage of the financial embarrassment of many planters during these years of low sugar prices to buy their land (Lasserre 1961: vol. 1, 390–1, 403–5; Pluchon 1982: 419–20; Revert 1949: 371; Schnakenbourg 1980: 212–20; Schnakenbourg 1984: 88–91).

The sugar industry in Guadeloupe by the early twentieth century was owned by French investors, as indeed English investors owned the industry on St. Kitts and Antigua; but on Martinique and Barbados the planter class remained in control. Why this contrast? The answer would appear to lie in the fact that Martinique and Barbados were respectively the regional centers of the French and English Lesser Antilles and the entrepôt trade each had carried on with other islands over the centuries had permitted the accumulation of reserves of capital. Martinican and Barbadian planters were thus able to finance the introduction of the central factory system without resort to metropolitan funds (Schnakenbourg 1984: 91).

Northeast Brazil shows yet another pattern of change among the old sugar colonies in response to the crisis of the nineteenth century, but it too belongs to the group of producers that made a slow and gradual increase in production (Fig. 7.2). Among the significant sugar-producing states of the northeast – Bahia, Sergipe, Alagoas, Pernambuco, Paraiba – Pernambuco

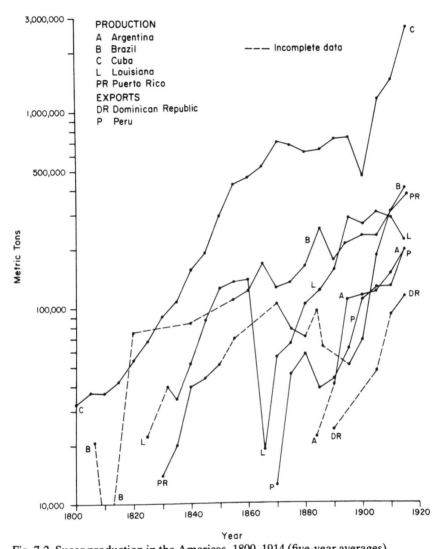

Fig. 7.2 Sugar production in the Americas, 1800–1914 (five-year averages)

Sources: Argentina: Guy 1984: 151. Brazil: Deerr 1949–50: vol. 1, 112–13. Cuba: Deerr 1949–50: vol. 1, 131. Dominican Republic: Cassá 1979–80: vol. 2, 136, 153. Louisiana: Deerr 1949–50: vol. 1, 126. Peru: Albert 1976: 13a, 14a. Puerto Rico: Scarano 1984: 8 (for 1828–52); Deerr 1949–50: vol. 1, 126 (for 1852–1915).

emerged as the most important with an increase in production from about 10,000 tons of sugar per annum at the beginning of the century to about 50,000 tons by mid-century and 150,000 at the end (Fig. 7.2). The increase was due in part to the same factors operating in the Caribbean: improvements in the mills, the introduction of new varieties of cane, and the fact that many of the planters at last began to manure their fields, but the industry also expanded. This came in part through an intensification of land use. There were, for instance, about fifty plantations in the coastal Pernambucan parish of Serinhaém at the turn of the nineteenth century and ninety-six in 1869, despite the fact that the parish had been reduced in size through the dismemberment of territory to found new parishes. The sugar industry was also a force behind the westward movement of the frontier. As a consequence of this activity, the 300 plantations in Pernambuco at the end of the eighteenth century had become 1,300 by the 1870s when the central factory system began to be introduced, and at this time there were between 3,000 and 4,000 plantations in all of northeast Brazil (Galloway 1968: 292–6; 1971: 591).

The expansion in the industry was accomplished without a major crisis in the supply of labor. Indeed, for the planters of northeast Brazil abolition represented a financial, political and emotional problem, but not a labor problem. A number of factors explain this circumstance. One was the lateness of abolition in Brazil – on May 13, 1888 – and it came after a series of anti-slavery measures, beginning in 1830 with the Anglo-Brazilian Treaty against the slave trade, that had given planters notice of the impending end of slavery, and had allowed them to make preparations. A second factor was the existence in the region of a free rural population. While there was indeed some discussion of encouraging the immigration of indentured workers, the planters turned to the local source of labor and had already begun to recruit on a significant scale even by mid-century. Many of the rural free in fact already lived on the plantations where they were known as *moradores* (from the Portuguese *morar* to live) and formed a numerous tenant class. They were retainers in a patriarchal society who performed errands and small duties for the planters, and kept an eye on the planters' property in return for permission to build a cabin and use a patch of land for the cultivation of food crops. With the decline of slavery, the planters required the *moradores* to work some days a week in the cane fields in return for their plot of land. Many of the ex-slaves on abolition simply became *moradores*, although not always on the same plantation where they had been slaves. The abundance of land still affected the behavior of the planters: they still had so much that they were able to exchange the use of land for labor and as they owned virtually all the land in the cane-growing region, they were able to deny access to this land to former slaves except on their terms. A slow transition characterized the move from slavery to freedom on the plantations of northeast Brazil, and even so that freedom as in the Caribbean was severely constrained (Galloway 1971).

The transfer to the central factory system in the northeast was a slow process that began only with government encouragement in the mid-1870s when both provincial and national governments offered to guarantee a minimum annual rate of interest on capital invested in central factories. The guarantees attracted some foreign interest and in particular two British companies, Central Sugar Factories of Brazil Ltd. and North Brazilian Sugar Factories Ltd., undertook to build factories. The first of these companies did succeed putting into operation four factories before going into liquidation in 1886; the second company achieved even less, only building one factory before disappearing from the scene. Managerial incompetence and poor machinery were major factors in the failures, and the very low price of sugar certainly did not make matters any easier, but a fundamental problem was the strategy of separating the ownership of processing from ownership of cultivation. The companies had difficulty in contracting for sufficient quantities of cane at prices that would permit them to make a profit while the planters still had their own mills and could make sugar in the traditional way if they did not like the contracts the factories offered. The remedy lay in reuniting the ownership of land and factory, a lesson that was soon learnt, and henceforth those who invested in the modernization of the industry took care to acquire lands for their central factories. A central factory with its own lands became known in northeast Brazil as a *usina* and its owner as a *usineiro*. The increase in the numbers of *usinas* led, of course, to a further concentration in landownership as planters sold their fields to the factory owners and withdrew from the industry. Transportation was a key factor in the diffusion, *usinas* being built first along the lines of the railways or along the coast so as to have easy access to the ports. By the end of the century, rather more than half of the sugar in Pernambuco was being processed in *usinas*, but even by 1920 the transition to the new system was by no means complete (Galloway 1968; Eisenberg 1974: 85–118).

As elsewhere, in the northeast of Brazil the modernization of the sugar industry produced a transformation in the structure of society: slaves disappeared as a social category, the *morador* class became more numerous while its conditions of tenancy were harshened by the requirement to provide labor, the *lavrador* class merged with those planters who stopped grinding their own cane and sent it instead to *usinas*, and many landowners sold out to *usineiros*. The *usineiros* formed a new elite made up almost entirely of planters who had managed to make the step to ownership of *usinas* through forming associations with neighbors which in turn raised money with the aid of government guarantees of interest on their loans. The sugar industry of northeast Brazil was not an attractive investment for American or European capital, and the industry remained in very large part Brazilian-owned.

Ironically, perhaps, the *usinas* did not help the northeast preserve its European markets where its sugar rapidly gave way to the competition from beet. The United States, with a strong refining industry of its own, wanted the muscovado produced by the traditional mills, not central factory sugar, and found suppliers among the Caribbean producers. Argentina, once a market, took less Brazilian sugar as it developed its own industry. With competition making the export of sugar increasingly difficult, the northeast sent more and more of its sugar to the south of Brazil where the population was increasing rapidly, began to market rum more aggressively and to look at other ways of expanding the market for sugar cane and its by-products. Interestingly, in the late nineteenth century, the *usineiros* were experimenting with the use of alcohol as a fuel – a harbinger of Brazil's response to the oil crises of the 1970s (Eisenberg 1974: 14–31).

Cuba and the new Caribbean producers

Against the record of the old Caribbean colonies during the nineteenth century, the success of five new producers stands out in strong contrast. At the beginning of the century, Cuba, Puerto Rico, the Dominican Republic and Trinidad were neglected Spanish colonies and the stretch of the Guiana coast that was to become British Guiana was still ruled by the Dutch. Britain acquired Trinidad and British Guiana during the Napoleonic Wars. The Dominican Republic experienced two interludes of independence, a Haitian occupation that lasted from 1822 to 1844, a brief return to Spanish rule and an attempt to join the United States before finally establishing a separate national existence in 1865. Cuba and Puerto Rico remained Spanish until 1898, Cuba thereafter becoming an independent republic and Puerto Rico an American possession. In all five territories there was a large increase in sugar production (Figs. 7.1 and 7.2). The sugar exports of Cuba increased more than 3,000% between 1800 and 1914 compared, for instance, to the increase of 400% in Barbadian exports over the same period and the decline in Jamaica. The token exports of Trinidad in 1800 rose during the century to over 50,000 tons, but the revival of the Dominican industry was even more dramatic: from the export of a mere 3,500 tons in 1875 to 120,000 in 1916. The extraordinary growth of the industry in these new colonies calls for some explanation.

One advantage shared by all the new colonies was an abundance of natural resources. There were extensive forests – at least in the early years of the century – to provide fuel and building timber, and the soils still had their natural fertility. Planters initially did not have to practice the labor-intensive, conservationist techniques that were essential elsewhere in the Caribbean. With higher labor productivity, they were better able to reward and therefore retain their labor supply. The persistence of slavery in Puerto Rico and Cuba until 1873 and 1886 respectively gave planters in these two islands an additional competitive edge over those who had been deprived of their slaves. All five colonies were sparsely populated at the beginning of the nineteenth century, and each benefited from an open frontier into which the industry could expand as the demand for sugar increased. The relative cheapness of land meant that the start-up costs of a plantation were comparatively low, as indeed was the cost of assembling land to supply a central factory. In 1913, when the sugar cane industry was making a come-back after the removal of the subsidies on beet, land suitable for sugar cane cultivation in the Dominican Republic sold for 12s to £1 per acre (Bryan 1978: 85). In Barbados – perhaps at the other extreme of the price range – even at the nadir of the industry's fortunes in the 1880s, plantation land sold for just above £20 per acre (Marshall 1975: 92). Hardly surprisingly, foreign capital flowed to the new producers in preference to the old, and American capital in particular went to Cuba, Puerto Rico and the Dominican Republic, all comfortably close to its homeland.

Cuba was the first and most important of these new sugar colonies. Its sugar industry, which had already begun to grow in the second half of the eighteenth century, was the major beneficiary of the collapse of sugar production in St. Domingue. On hearing of the slave revolt, the Cuban planters quickly set aside any anxieties that such a thing might overtake them, and began to exploit the opportunities that events had created. Indeed, until after the late 1830s, when technical innovations required large infusions of capital, the long-established Cuban landed families were able to finance and manage the expansion of the sugar economy (Knight 1977: 249). In January 1792, a Cuban landowner and noted Cuban economist, Francisco de Arango y Parreño (1952: vol. 1, 114–75), dispatched to the King a lengthy report on how the agriculture of Cuba might be developed. The Spanish government did make a constructive response, easing the flow of labor into the island by permitting slavers of all nationalities to bring their cargoes to Cuba and by reducing the duties on slaves. It also abolished import duties on machinery, even of foreign manufacture, a step which enabled planters to follow more cheaply the advice they were receiving from St. Domingue refugees on how best to modernize their mills and purging houses. Finally, in 1818, it at last removed all other legal impediments on trade between Cubans and foreigners (Parry and Sherlock 1956: 223–5).

The results of this encouragement – and the removal from the trade of the most important competitor – can be seen in the increase in sugar production from about 32,000 tons in 1800 to nearly 100,000 in 1830 (Fig. 7.2). The number of mills increased four-fold, from 245 in 1792 to about 1,000 in 1827 (Marrero y Artiles 1972–84: vol. 10, 151, 163). The authorities

(Fraginals 1976; Marrero y Artiles 1972–84; Thomas 1971) agree that the Cubans were building larger mills, although they are not entirely consistent in their statistics. Thomas (1971: 62) adapts the figures of Fraginals, and concludes that the average annual capacity of the mills and purging houses was as follows: 1761 – 40 tons; 1792 – 55 tons; 1804 – 130 tons; 1827 – 170 tons. French technicians from St. Domingue worked in all the larger mills and purging houses, and among other improvements introduced the Jamaica train as well as a far greater use of water power. Arango y Parreño was responsible in 1797 for installing the first steam mill, but the experiment was not a success and steam power did not become significant for another twenty-five years (Thomas 1971: 78–9). The introduction of Otaheiti cane, also via St. Domingue, made in Cuba, as elsewhere, a significant contribution to the increase in yields.

The demand for sugar combined with access to markets had in fact justified the investment of capital to bring the Cuban industry by the 1820s up to the highest technical standards of the time. For the next forty years, until the outbreak of the disastrous Ten Years' War in 1868, the expansion of the industry continued, and the Cuban sugar industry enjoyed what both Thomas (1971: 109–27) and Marrero y Artiles (1972–84: vol. 10, 163–208) refer to as its Golden Age. There was still abundant virgin land to draw into the industry, particularly in the center and east of the island, and by the 1860s the provinces of Matanzas and Las Villas had by far superseded in importance the original nucleus of the industry around Havana. The geographical extension of the industry was helped by the arrival of one of the most important products of the industrial revolution – the railway. The planters sponsored and controlled the railway companies, and they ensured that the development of the network of lines reflected their interests. As early as 1838, 45 miles of track were open to traffic, and by 1860, there were about 400 miles of track. Nor did labor present a problem to the planters during this period. Despite the illegality of the slave trade after 1817, they continued to be able to import large numbers of slaves, perhaps as many as 400,000 between the early 1820s and the end of the traffic in 1865. And even as the slave trade declined, the planters successfully turned to China, recruiting more than 120,000 workers there between 1843 and 1873 (Thomas 1971: 1532–3, 1541). Mills became more efficient as planters adopted steam power and replaced their traditional three-roller mill with horizontal presses. The wealthier planters began to buy vacuum pans in the 1830s and centrifugals after 1850. At mid-century, therefore, the Cuban industry exhibited quite a variety of technology, from mills little changed in style from those of the early eighteenth century to those that embodied the latest in industrial engineering. The capacity likewise varied enormously, with some mills producing no more than 130 tons a year, to the modern ones that produced over 1,000 tons. There were about 1,500 mills in Cuba, and together they accounted for about 25% of all sugar produced in the world, beet sugar included.

The Ten Years' War (1868–78) brought a great deal of damage to the sugar industry, particularly to mills in the eastern and central districts of the island, and although it marked the end of the Golden Age it was really only one of several contributing causes of the change in the pattern of development. In the early 1860s, the sugar industry was on the threshold of the massive reorganization, economic as well as social, that the introduction of the central factory system entailed. Its introduction coincided with abolition and the end of Chinese indentured labor. The price of sugar was declining and markets uncertain. Cuba had had since 1846 access to the British market, but there competition from beet was increasing, and meanwhile the new sugar industry in Louisiana raised uncertainties about the future of Cuban sales in the United States. With the return of peace in 1878, the sugar industry resumed its growth, but in very different circumstances.

Slavery came to an end gradually. There were still over 350,000 slaves in the 1860s, but when the Spanish government abolished slavery in 1880 there were about 200,000 slaves on the island. The government did not grant immediate freedom, but instead insisted on a further eight-year period of bondage (the *patronato*) during which, however, the slaves could buy their release. The stated reason for this delay was to minimize the risk of social upheaval and disruption to the economy, but another explanation is that an impecunius government unable to pay an indemnity chose instead to mollify owners by allowing them the use of slave labor for eight more years. Rich planters, more able than poor planters to turn to other forms of labor, sold or freed their slaves, while the poor clung to theirs, but *patronato* disintegrated, and in 1886, with forced laborers down to only 33,000, the government abolished it (Corwin 1967: 305–11).

During the last third of the nineteenth century a new system of organizing the cultivation of cane arose in Cuba, the so-called *colonato* which was a response to the end of slavery and of Chinese indentured labor as well as a response to the central factory system. The *colonato* included two categories of *colonos*: tenants who leased land on condition they cultivated sugar cane to send to the landowner's mill or central factory and landowners who contracted to supply their cane to a mill or more usually with the passage of time to a central factory. Within each category there was a range from the comparatively wealthy to the poor. By leasing land to *colonos* a mill or factory owner could increase the area of cane cultivated on his estate without incurring additional labor costs to himself, and the increased production could justify the capital expense of technological improvements in milling and manufacturing. Some slaves were induced to become tenant *colonos*, cultivating sugar cane on small plots with their own family labor; other tenants took larger plots and contracted for laborers as best they could. Among early recruits of the wealthier kind to the *colonato* were plantation owners whose mills and purging houses had been destroyed during the Ten Years' War and who lacked the capital to rebuild. The expansion of the central factory system brought many more former mill owners into the *colonato*, as the following figures show: in 1878 there were still 1,190 mills in Cuba, 850 in 1891, but only 207 in 1899 although this last figure reflects not only the progress of centralization but destruction in the War of Independence (Lopez 1982: 76). Theoretically a landowning *colono* was able to negotiate with factory owners over payment for cane but in fact only those *colonos* whose land was within easy access to two or more factories were in a position to do so. A factory could bind a landowning *colono* to his factory by loaning equipment and by advancing cash to finance the cultivation of a crop.

The relationships in the *colonato* between cultivators and processors of sugar cane were in fact hardly original, having counterparts in other canegrowing regions of the Americas and elsewhere. In Cuba, however, the *colonato* rapidly took on great importance, perhaps because the Ten Years' War, abolition and the introduction of the central factory system came so closely together. *Colonos* first appeared in significant numbers during the 1870s and already by "1887 a third to two-fifths of Cuban sugar was grown by the *colono* system" (Thomas 1971: 276). As a means of organizing the cultivation of cane, the *colonato* was a success, accommodating the needs of central factory owners, mill owners unable to repair war damage or unable to finance new equipment as well as owners of a few acres who wished to participate in the cultivation of a cash crop, but it did not attract many former slaves for reasons that are not yet clear. The 1899 census revealed that colored renters and landowners amounted only to a modest proportion of the total, and that their land comprised only 4.5% of the area in sugar. The members of the ex-slave class worked as hired hands, grew provision crops on small-holdings or moved to the cities.

The shortage of field labor remained a perennial problem that was particularly acute during periods of peak demand such as the harvest season. In some provinces the sugar industry so dominated landownership that the landless had either to work for what wages were offered or move elsewhere. In addition to denying the rural poor access to land, large landowners had other ways of manipulating the labor market. They could easily ensnare their laborers in debt by paying advances on wages, although debt peonage did not become widespread. Another strategy was to pay wages in scrip valid only in the estate stores. They persuaded the government to pass vagrancy laws against the so-called "idle" poor, and there is some evidence of planters employing armed men to coerce people to work for them. In the last years before independence in 1899, Spain encouraged emigration not only to provide Cuba with more labor but also to relieve population pressure at home. Many thousands left the poor regions of northwest Spain: for some it was a permanent move, for others an annual journey for the sugar harvest (Scott: 1984). But this source was insufficient and the labor demands of the expanding Cuban industry prompted yet another mass migration, this time from within the Caribbean. The 27,000 Haitians and 23,000 Jamaicans who came to work in the cane fields of Cuba during the First World War (Thomas 1971: 540) set a pattern of seasonal migration that was to last for many years.

The introduction of the central factory system changed the structure of land tenure in Cuba, as of course it did elsewhere, but it also changed the finances of the Cuban industry. The building of factories, complete with narrow-gauge railways to carry the cane from the fields to the factories, and the expansion of the island's rail network called for investments on a scale not before seen in Cuba. Indeed, the modernization of the industry was a matter of great urgency, for the competition from beet combined with the great decline in the price of sugar after 1884 meant it had to be extremely efficient if it were to retain a profitable share of the market. The capital requirements to effect the change were beyond the resources of the Cuban elite, weakened as it had been by the Ten Years' War and then the War of Independence, and ownership of the industry passed into foreign, especially American, hands.

The consumption of sugar in the United States increased rapidly during the late nineteenth century, and various groups striving to supply this market sought to influence government policy to their own advantage. The Louisiana and Hawaiian industries had their spokesmen as did a nascent beet industry, but Cuba, so conveniently close, was an obvious source of supply and already by 1880 had a large share of the U.S. market. Support for American investment in Cuba was a predictable development, part of the wider United States interest in the Caribbean and Central America that led to strong commercial penetration of the economies of most of the countries, armed interventions in the internal affairs of a few and, in the case of Puerto Rico, annexation. The McKinley tariff reforms, in force between 1891 and 1894, which were in part at least the result of lobbying by Americans with Cuban investments, removed duties on raw sugar and molasses and helped push sugar production in Cuba to 1 million tons. Dependence on the American market brought Cuban producers into conflict with the American refiners, organized after 1888 into the powerful American Sugar Refining Company – or Sugar Trust – which sought to keep prices low and to import poor grades of sugar (Thomas 1971: 288–9).

American investment in the industry continued during and after the American occupation at the time of Cuban independence. Large agricultural production companies such as United Fruit became prominent participants.

In a new twist in the international organization of the sugar trade, major industrial consumers in moves to integrate their activities vertically began to secure their own supplies of sugar by buying lands and factories. Hershey's, of chocolate bar fame, and the Charles Hire Company, manufacturers of root beer, bought factories in Cuba. The refiner, Havemeyer, organizer of the Sugar Trust, had bought into the Cuban industry as early as 1892. Some international sugar brokers whose business it was to buy sugar for resale to refiners and industrial consumers around the world took steps to secure their own supplies of cane. In 1915, the managing partner of the New York office of the brokers Czarnikow Ltd. of London, co-founded the Cuban Cane Sugar Corporation. The extent of American ownership is difficult to measure because of the distribution of shares, and often Cuban owners lived in the United States. In 1919, one estimate claimed Americans owned between 40% and 50% of all factories; another placed the figure at 35%, but the factories accounted for 50% of the sugar Cuba produced. Other factories were owned by Canadian, Spanish and other national interests (Thomas 1971: 536–43).

By the end of the First World War, the Cuban sugar industry was preeminent in the world, producing annually about 4 million tons in 200 central large factories. It had overcome triumphantly, it would seem, the challenges of abolition, technological innovation and the competition from beet. Viewed from another perspective, however, there was a high pyrrhic element in the story of success. The centralization of the industry had led to an even greater concentration of landownership and wealth, and the ownership of the industry had passed into foreign hands. These two consequences of modernization did not auger well for the social and political stability of the island. The dominant role the industry had assumed in Cuba meant that the island's economy had become even more vulnerable than before to the fluctuations in the international price of sugar. Successive Cuban governments have been unable to overcome this debilitating dependence.

Puerto Rico, the second Spanish Caribbean colony to develop significant exports of sugar during the nineteenth century, has a special place in the historical geography of the sugar cane industry because here, for the last time since the settlement of São Tomé three centuries before, planters attempted to base their activities on African slave labor. Puerto Rico is a far smaller island than Cuba, and never came close to achieving Cuba's importance in the international sugar trade, but even so, by the 1870s, it had come to rank second among Caribbean exporters. Shortage of capital was a basic factor determining the character of the industry: it helps explain the late start, the limited the number of slaves the planters could buy and the serious delay during the second half of the nineteenth century in the adoption of technological innovations. Only after the American annexation in 1899 did this problem disappear, and then the sugar industry flourished (Fig. 7.2).

The Spaniards had brought sugar cane to Puerto Rico early in the 1500s, but for the following three centuries Spain ignored the economic potential of the island, and the little amount of sugar produced was only for local consumption. As a consequence of this neglect, there was only a small merchant community on the island at the end of the eighteenth century and it had neither the capital nor the technology to take advantage of the demand for new suppliers of sugar that the slave revolt in St. Domingue created. Spanish and foreign capital flowed in preference to Cuba which not only had a well-established industry with room to expand but had the further advantage over Puerto Rico of greater proximity to the United States market. It was not until about 1820 that Puerto Rico began to develop an export industry. For this change, the commercial reforms Spain prescribed in 1815 were a helpful factor, but more important was the British occupation of the prosperous Danish free port of St. Thomas between 1807 and 1815. One effect of the occupation was to restrict greatly the wide-ranging contacts the merchants of St. Thomas had developed over the years, but trade with nearby Puerto Rico was still possible and many of the merchants began to invest in agriculture there. This interest continued after the British restored St. Thomas to the Danes (Scarano 1984: 16–25). A further factor was the arrival in Puerto Rico of refugees from Haiti and from the Wars of Independence in Venezuela with both capital and knowledge of sugar technology (Gil-Bermejo Garcia 1970: 129).

The development of a sugar industry changed the economic and social structure of Puerto Rico. While sugar plantations were by 1830 widely distributed around the coast, cultivation soon became concentrated in the fertile coastal districts of the south and west, shifting thereby the main locus of economic activity in the island away from San Juan on the north coast. Mayagüez and Ponce, the towns that benefited most from the new industry, were by 1850 centers of commercial and political influence. The island population, when these changes were set in motion, was largely a peasant one, of mixed racial origin, raising cattle and cultivating food crops for its own use as well as for sale to the military in San Juan, to the crews of passing ships and with some export in good years to other islands. As sugar grew in importance, the role of the peasantry in the island's economy declined. Peasants left the new plantation zones to reestablish themselves in the interior upland regions that were not attractive to the sugar industry. The fact that this escape route existed for the peasantry meant that sugar planters had difficulty in recruiting free labor and so turned instead to slavery. The slave population grew from 18,616 in 1815 to 41,818 in 1834, and while free labor did work in the sugar industry, slaves formed the core of the work force on many plantations until abolition in 1873 (Scarano 1984: 3–34).

The steady growth of the industry faltered in the years after 1870, and by 1900 exports had fallen back to the levels of the 1850s. The competition from beet sugar and the low sugar prices during the last decades of the nineteenth century were, of course, major problems, but the failure of Puerto Ricans to adopt such important innovations as central factories and railways reflects the undercapitalization of the industry. The contrasts with Cuba became very marked. In 1899, the 345 Puerto Rican mills, half still worked by oxen, had a collective refining capacity equal to one tenth of that of the 207 Cuban mills and factories. Central factories in Puerto Rico were a rarity and railways limited to a few miles of track. In one central factory, oxen and not locomotives hauled the carts along the 3 miles of private track. The industry, clearly, was in crisis, a fact reflected in its declining contribution to the Puerto Rican economy – it accounted for only 21.6% of all exports in 1897 compared to 68.5% in 1871 – and might soon have been reduced to the role of producer of poor quality sugar for the local market had it not been for the American annexation in 1898 (Bergad 1978: 64–7).

The U.S.A. had throughout the nineteenth century been an important market for Puerto Rican sugar, but the inclusion of the island, in 1901, within the customs area of the U.S.A. was to bind the industry to the American market. Puerto Rican sugar now entered this market duty free which meant in fact that it was subsidized by the amount of the duty foreign producers had to pay on the sugar they sold in the U.S.A. American investors may well have anticipated this extension of tariff protection to the island because already by 1898 they were buying large tracts of land. Three large American companies – to be joined by a fourth in 1926 – became the major participants in the Puerto Rican industry and, in a repeat of trends in Cuba, a once largely domestically owned industry passed into foreign hands. Modernization proceeded so rapidly that by 1910 there were forty-one central factories. The companies derived much of their cane from their own vast acreages which they cultivated with wage labor, paying sufficiently well as to attract workers down from small-holdings in the interior, and an important percentage of the total crop was grown by *colonos*. As in Cuba, the *colonato* was a diverse social group composed of renters and sharecroppers as well as of owners, some of whom at least may once have had their own small mills. The effects of this reorganization of the industry and the capital infusion soon showed results in a rising tide of exports as Puerto Rico evolved into a modern sugar colony (Bergad 1978: 77–81; Mattei 1984).

The Dominican Republic, formerly Hispaniola, provides a third example of a new sugar industry developing in a neglected Spanish Caribbean territory largely through the commercial and political intervention of the United

States. The modern Dominican industry began in the 1870s, rather late compared to the industries in Cuba and Puerto Rico. It then grew rather slowly during the next decades and contributed relatively little to international trade, the exports rising from 7,000 tons in 1880, to 48,105 in 1905 and 101,428 in 1914. The demand for cane sugar during the First World War gave a boost to Dominican producers, but American occupation of the country from 1916 to 1922 with its promise of stable government further encouraged investors. By 1925, exports had reached 300,000 tons, and the transformation of a sparsely populated backwater into a sugar colony was well underway (Cassá 1979–80: vol. 2, 136, 153).

Sugar was, of course, not a new crop on Hispaniola: it had been grown there since the beginning of Spanish settlement, first as an export crop, and, after the collapse of this industry in the early seventeenth century, it continued to be grown on a very modest scale for local use. During the seventeenth century, the inhabitants had lived mainly from subsistence cultivation assisted by a very modest export of hides to Spain. In the 1700s, Hispaniola returned to the sugar industry, but in the secondary role of supplier of beef and beasts of burden to the French plantations in St. Domingue (Cassá 1979–80: vol. 1, 165, 170). Indeed, Spanish Hispaniola became a peripheral part of the French imperial system and stood in much the same relation to the French plantations in St. Domingue as the cattle ranches in the interior of northeast Brazil stood to the plantations of the coast. This functional link between the two parts of the island was severed by the slave revolt in St. Domingue, and Spanish Hispaniola lapsed once more into economic stagnation.

In the early nineteenth century, Hispaniola – the Dominican Republic – had, like Cuba and Puerto Rico, the advantages of fertile soils, extensive areas of unused or underused lands and proximity to the large market of the United States, but the country also had the disadvantage of political instability that lasted throughout the nineteenth century, and this was the basic reason would-be investors in the sugar industry made their investments elsewhere. In 1795, Spain ceded Hispaniola to France, but the French were never able to impose a stable government on the various warring factions and interest groups, and in 1808 a coup in the city of Santo Domingo restored the country to Spanish rule. This proved very unsatisfactory, and in November 1821, as the result of a new coup, the country became independent, but in February 1822, before this independence could be consolidated, the Haitians invaded and remained until 1844. The misrule during this occupation led to the emigration of many of the Spanish-speaking elite, fields and pasture reverted to forest, and the population declined to about 60,000, half of what it had been in 1800. In 1844, a successful coup led to a second declaration of independence, and the country finally took the name of the Dominican Republic, but government degenerated into coups and counter-coups, punctuated only by new invasions from Haiti. A voluntary return to Spanish rule in 1861 pleased neither Spain nor the Dominicans, and in 1865 the country became independent for the third time. Dictatorship or near anarchy continued to characterize Dominican politics with one tyrant, Ulises Heureaux, providing a measure of stability between 1882 and his assassination in 1899. The U.S.A. refused an offer from the Dominicans to annex their republic in 1865, but, nevertheless, in concert with its interventionist policy towards the small states of the Caribbean and Central America, it did become more active in Dominican affairs, eventually taking over in January 1905 the administration of the Dominican customs in an attempt to restore some order to the Republic's finances, and, in 1916, landing the marines to occupy the country (Fagg 1965: 145–54).

It is hardly surprising, given this woeful record, that the Dominican Republic remained for so long something of an economic wasteland, that foreign investors shunned it and the Dominicans were unable to accumulate capital of their own. Indeed, the Republic owes the beginnings of its modern sugar industry not to the initiatives of one of its governments, but to the turmoil elsewhere. The onset of the Ten Years' War in Cuba (1868–78) caused some sugar planters to emigrate from Cuba to the Dominican Republic. These planters had both capital and technological knowledge. They bought land, built new factories, and found that good wages attracted a sufficient supply of labor from among the local peasantry. They were also able to contract with landowners near their factories to supply cane. In other words, they introduced the Cuban *colono* system. The modest success of these refugees in establishing sugar plantations was reflected in the appearance of sugar as a Dominican export by the early 1880s (Bryan 1978).

This achievement led other foreigners – American, British and even Italian – to invest in the expanding industry. A powerful attraction for these investors was the cheapness of land in the Dominican Republic compared to other Caribbean sugar colonies. Land suitable for sugar cane cultivation could be bought for between 4s and £1 10s per acre in 1880 when it cost £7 per acre in Cuba in 1863, and even £3 5s in the depressed island of Jamaica in the 1860s. Even by 1913, good land in the Dominican Republic still sold for between 12s and £1 per acre (Bryan 1978: 84–5). The region the investors particularly favored was the southeast of the country, around the town of San Pedro de Marcorís, where there were fertile soils and good coastal communications. By the 1890s, over half of the Dominican production came from this region, and it maintained its predominance until after the First World War. The construction of railways during the presidency of Ulises Heureaux and further improvements in roads and railways during the years of the American military occupation greatly assisted the expansion of the industry. The highly competitive international market gave the commercial advantage to well-capitalized companies that were able to install the most cost-efficient machinery, and so these years leading up to 1914 saw a consolidation in ownership of the Dominican industry as the more successful companies grew at the expense of the failing. This meant in effect foreign-financed companies took over a larger and larger share of the industry as few Dominicans had the capital resources to modernize factories. These same well-financed companies could also respond most effectively to opportunities brought both by the higher sugar prices during the First World War and by the administrative stability of the American occupation, and so the process of consolidation continued. By 1925, one American company, the West Indies Sugar Company (later the Cuba Dominican Sugar Company), accounted for 40% of Dominican production, a second American company, the South Porto Rico Sugar Company, for a further 21%, and the Italian Vicini Group for 11%. Most of the remaining producers were subordinated in various ways to the West Indies Sugar Company (Cassá 1979–80: vol. 2, 131–6).

The expansion of the industry brought an increased demand for labor, but that labor had to be cheap. In response to the decline in the international price for sugar during the late nineteenth century, planters lowered the wage rates with the result that many workers who still owned their own plots of land returned to full-time peasant cultivation. Dominicans continued to work in the few more skilled occupations on the plantations, but for field labor, and in particular to meet the needs of the harvest, the industry's owners turned to immigrant laborers who, because they did not own their own plots of land in the Republic would not be in a position to bargain strongly over wage rates. They looked first to the Spanish-speaking Canary Islands and when these failed to provide a sufficient flow of manpower, they turned for laborers to Puerto Rico, to the nearby Virgin Islands, to St. Kitts, Nevis, Anguilla, Antigua and even to the feared Haiti (Castillo and Cordero 1980: 23–5). Many of these workers only came for the harvest and so seasonal inter-island migration of cane-cutters became part of the Caribbean scene.

The expansion of the sugar industry in Trinidad and Guiana, two colonies acquired by Britain in the aftermath of the Napoleonic Wars, stands in marked contrast to the decline of the industry in some of the old British colonies and its precarious survival in others. Trinidad had experienced little agricultural development during the centuries of Spanish administration, and in the early 1800s its sugar exports amounted to only a few hundred tons annually, but it soon became as important a producer as Barbados. In Guiana, during the eighteenth century, planters of Dutch, French and English origin had cultivated sugar cane on the low-lying coastal plain behind the protection of a sea wall, but their contribution to the sugar trade was still modest by the time they came under British rule. Yet, by the mid-nineteenth century, Guiana had become the major exporter of sugar in the British Caribbean (Fig. 7.1). The explanation for this turn of events in these two colonies is to be found in a combination of favorable circumstances that included a successful solution from the planters' point of view of the labor problem and an inflow of capital that permitted the adoption of technological innovations. A consequence was the more rapid modernization of the industry than in the other British Caribbean colonies.

In the decades between the end of apprenticeship in 1838 and the beginning of the First World War, two themes characterized the policy of the planter class of Guiana and Trinidad in its labor relations: first and foremost, they were persistent in their struggle to secure a cheap and reliable supply of labor and, secondly, they were willing to experiment with new labor relations and new sources of labor. An initial difficulty in 1838 followed from the fact that both colonies were sparsely populated, with unoccupied crown land in close proximity to the plantations onto which the former slaves could move to settle as squatters. In Guiana, they could, if they so wished, filter well away from the coast, inland up the many rivers beyond reach of any restraining force. Movement was much more constricted on the island of Trinidad, and the planters hoped to limit this loss of laborers by persuading the authorities to distribute the crown land only in large acreages rather than in small peasant holdings. Such a strategy, however, was not successful for ultimately the prevention of migration onto the unoccupied lands and the eviction of squatters already there were difficult, disagreeable policies to put into effect, and unrealizable without a large and expensive apparatus of coercion. The planters in both colonies had to rely on inducements to retain laborers, and in this they were aided by the ambiguous position the former slaves found themselves in. While the former slaves might have wished to be free of the plantations once and for all, squatting on a small-holding in the interior was unlikely to provide an existence above the subsistence level; but working for the planters who now paid wages had the attraction of entry into the consumer economy, albeit on a modest scale.

Some of the former slaves did indeed choose to remain on the plantations, working in the cane fields for a wage, living in houses and cultivating provision plots allocated to them by the planters. Others settled close to the plantations, even on abandoned plantation lands, so that they could easily work for the planters. In Trinidad, to attract laborers to the fields, planters adopted the system of playing by the "task" rather than by the day, except in the harvest season. A "task" was a measure of work said to equal what a slave had accomplished in a day. Yet a worker could finish a "task" in four to five hours, and so complete two a day if he had the energy. The pay for a "task" rose from 30 cents to 50 by June of 1840. These were the highest wages in the British Caribbean and compared very unfavorably from the planter's point of view to the 20 cents a day paid in Barbados where the former slaves had no alternative to estate labor (Wood 1968: 52–3, 59). Moreover, even with these high wages – more than some planters could bear – a reliable labor force proved elusive as workers took time off from the cane fields to cultivate their provision grounds and to make trips to town to sell what they had grown. The numbers employed in the industry declined, some plantations ceased production, and exports decreased slightly but steadily from year to year. The industry in Guiana was in even more serious difficulties, as the exports reflected: in the early 1840s they were 43% lower than in the years immediately before abolition. It was clear to the planters and both colonial governments that a new strategy was in order if the decline in the industry was to be reversed (Adamson 1972: 34–56; Williams 1964: 86–101; Wood 1968: 46–55).

The answer lay in immigration and, specifically it turned out, in immigration of indentured laborers. There was a nearby source of labor in the small British islands in the Leeward and Windward groups, where, after the end of apprenticeship in 1838, there were willing recruits attracted by the high wages on offer in Trinidad and Guiana. Local authorities objected to this loss of laborers, but the movement of free people was difficult to stop if inter-island schooners were willing to transport them. In 1844, the Colonial Office did order Trinidad to stop paying a bounty to recruiters for each laborer they brought to the island, but, nevertheless, the migration was considerable. Between 1839 and 1849, 10,278 immigrants from the islands entered Trinidad and 7,582 entered British Guiana. By contrast, only 790 went to Jamaica. It is not known how many of these arrivals were permanent immigrants, and the Barbadians in particular maintained a strong identification with their home island. Yet whether Barbadian, Antiguan, St. Kittian or from some other island, the immigrant former slaves behaved much as the Trinidadian and Guianan former slaves did, working on or off the plantations as they saw fit. In Trinidad, Barbadians found their way into the lower ranks of the police force – not a move to make them popular (Wood 1968: 65–6).

The role of a second immigrant group – the Portuguese from Madeira – evolved in a way that was not originally intended, but yet in the long term served very well the interests of the planters. Madeira had long been a familiar name to the English. To the general public, it was the source of a very drinkable dessert wine; sailors also knew the island as the last European port of call outward bound on voyages to the tropics, and ever since its use as a place of rest and recuperation for officers wounded in the Peninsular War it had been a winter resort for the upper classes. Many a West Indian planter had broken the long journey back from England with a short stay on the island. The wine economy that had supported Madeira during the two and a half centuries since the collapse of its own sugar industry was by the 1840s in deep trouble. Tastes in wine drinking had changed so that Madeiran wine no longer enjoyed the demand it once had had, and to make matters worse, in 1852, disease destroyed the vineyards. Emigration had long been a feature of Madeiran life, and now there was an extra impulse to leave. Understandably, the planters in Trinidad and Guiana looked forward to receiving laborers used to hard agricultural work, a low standard of living, and from a European source they knew well. Despite all blandishments, Trinidad was able to attract very few Madeirans, but 21,811 had gone to Guiana by 1861 (Wood 1968: 102). The Madeirans soon appreciated that plantation labor did not offer them good prospects for improving their economic situation, and so those who could manage it left the plantations. As in the case of the former slaves, many entered petty commerce.

Planters clearly recognized this avenue to upward mobility as a threat to the plantation labor supply, and in order to block it, as well as perhaps to raise some revenue, the Guianese authorities in 1842 increased the annual license fee from £1 to £2 for peddlers and from a nominal amount to £5 for shops. These were significant sums for the time and of course applied to the Portuguese as well as to the former slaves. The elite recognized the need for petty commerce in the colony, but preferred the Portuguese in the role and took steps to bring this about. One very effective method of discriminating in favor of the Portuguese was followed by the import-export merchants in Georgetown, the capital, who extended credit on easy terms to Portuguese traders while refusing credit to the former slaves. The results of this policy were soon apparent: by 1852 the Portuguese received nearly 70% of all shop licenses issued in the colony (Wagner 1977: 410–11). In Trinidad, too, the Portuguese came to dominate local retail trade at least until the 1860s when the Chinese offered competition (Wood 1968: 106–7). The Guianese elite was successful in using the Portuguese to block the upward mobility of the former slaves and, in effect, to deny them some of the economic opportunities they might have expected to follow from emancipation. Moreover, the elite now had between it and the black population a white immigrant group that was bound to become resented and would become the focus, rather than the elite, of popular discontent. Thus racial politics was another planter strategy intended to limit the options of the former slaves and keep them on the plantations. In Trinidad and Guiana, however, it was unsuccessful: not enough of the former slaves were willing to work on the plantations, and the planters turned to a third immigrant group (Moore 1975; Wagner 1977).

In Trinidad and Guiana, more so than anywhere else in the Caribbean, Indian immigrants provided the solution to the plantation labor problem. The first Indians arrived in 1838, but because of reports of their poor treatment on the plantations, the British government stopped recruitment between 1844 and 1848. From 1851, however, a stream of indentured workers continued to flow from India – more strongly in some years than in others – until 1917, and by then the Indians had not only helped the sugar industry to survive but also had transformed the cultural geography of the two colonies. Altogether, 145,000 Indians went to Trinidad (Williams 1964: 100) and 236,205 to Guiana (Adamson 1972: 46). Not all remained of course: some took advantage of their right to a return passage home but many stayed, so that by 1871 they accounted for 25% of the population of Trinidad (Wood 1968: 158). In addition to the reliability of this source of labor, the administrative and financial assistance the colonial and imperial governments gave the scheme represented a significant subsidy of the labor costs of the plantations.

Indentured Indian labor was one of the reasons, in the opinion of Beachey (1957: 102), for the success of the sugar industry of British Guiana compared to that of other British Caribbean colonies. The Indians did leave the plantations after their period of indenture was over, but the colonial government ensured that they would remain in the coastal strip and thus close to the plantations which could employ them as day laborers by enforcing a very restrictive policy towards the alienation of crown lands. Statistics provide a measure of the severity of this policy: between 1839 and 1890, the colonial government made only twenty-two grants of crown land, although during the 1890s it became in comparison lavish with its grants, making an average of thirty-seven per annum (Adamson 1972: 135–6, 259). Without industrial jobs to turn to, with entry into commercial life blocked by the Portuguese and without legal access to crown lands, the time-expired indentured workers settled into villages beside the cane fields, grew rice and other food crops and provided a reservoir of manpower for planters to draw on during the seasons of peak labor demand.

In Trinidad, the labor situation evolved somewhat differently and heralded attempts to organize the cultivation of cane on a more egalitarian basis, by small-holders rather than by planters, that were to become more common in the twentieth century. There was easier access to land in Trinidad both for blacks as well as Indians. Policy towards the alienation of crown land was less restrictive, abandoned estates were subdivided and some estate owners found it more profitable to rent their land in small plots than to continue as sugar planters. Tenants and small-holders who cultivated cane for delivery to factories were known as cane-farmers. Some factories rented land for cane-farming. This began in a very modest way during the late 1870s and early 1880s, grew in importance towards the end of the century and by 1906, some 11,500 cane-farmers – rather more than one half of whom were of East Indian origin – cultivated one third of the sugar cane delivered to the factories. The size of the cane-farmers' plots varied from a mere fraction of an acre to more than 100 acres, but most cultivated from 3 to 5 acres. There were several advantages in cane-farming for the small-holder. It was an entry into the cash economy that required little skill or capital, and there was an assured market. Moreover, it provided quick returns compared to cultivating the other export crop of the island, cacao: seed canes were ready for harvest within eighteen months, ratoon crops within a year, while cocoa trees took fifteen years to mature. There were advantages, too, for the planters and factory owners. They encouraged cane-farming because they expected the farmers and their families, settled near the plantations, to be a source of labor during the harvest. They also discovered that cane-farmers working for themselves and unencumbered with the overhead costs of a large operation could grow cane more cheaply than the plantations. Two factors, however, convinced the planters to maintain a large acreage of their own cane and not to concentrate entirely on the manufacturing side of the industry. First, they could not rely on the cane-farmers to ensure a steady flow of cane to the factories and so had to have their own supplies to draw on. Secondly, they did not wish to place themselves in a weak position vis-à-vis the cane-farmers when it came to bargaining over the price of cane as they would be in if they were dependent on them for cane. The plantocracy took care, in other words, not to give the cane-farmers economic independence (Johnson 1972). These relations of power and self-interest between factory and supplier are among the enduring realities of the modern sugar cane industry and have bedeviled attempts to break down the plantation structure in favor of a more egalitarian style of organization.

In addition to the favorable solution to the labor problem from the point of view of the planters, the second advantage that helps explain the growth of the sugar industry in Trinidad and Guiana was the ability of these colonies to attract capital to finance technological improvements. This was the case for several reasons. Immediately after emancipation in these colonies, as elsewhere in the British Caribbean, there were three possible sources of capital: the planters' own savings, advances from agents in Britain against the security of future crops and the funds planters received from parliament in compensation for the emancipation of the slaves. In fact, however, few planters had savings, and as the parliamentary money usually went to reduce debt there remained only the traditional line of credit to agents. Although the Colonial Bank of the West Indies was established in 1836, it was of little help to the industry as its charter prevented it from making loans on the security of land or buildings; the Bank of British Guiana had a similar restriction on its loan policies. However, Trinidad and Guiana had three attributes that made them the most promising places in the British Caribbean in which to invest: the fertile soils rather than worked-over lands, a solution to the labor problem and, very significantly, because of the newness of the industry, plantations unencumbered with debt. Plantations could be bought and sold without delays. When land prices dropped steeply following abolition – according to one estimate plantations in British Guiana in 1846 sold for 20% of their 1840 value – the purchasers, often from Britain, not only brought new capital but had the advantage of a very low cost of entry to the industry. A revitalized planter class must in itself have been an asset, with its new commercial contacts and a fresh outlook on the technological world of the nineteenth century (Adamson 1972: 164; Lobdell 1972: 36–8).

Not surprisingly, in comparison to the other British colonies innovations appeared early and were adopted rapidly on the plantations of Guiana and Trinidad. Guiana, in particular, was well to the fore. By mid-century, the industry had already converted to steam power in the sugar mills and was using steam pumps to drain the fields; by mid-century, also, twenty-five plantations had installed vacuum pans. John Gladstone installed the first one in British Guiana, on his Vreed en Hoop estate, in 1832, and the acceptance of this expensive equipment would have been even more rapid had not British duties discriminated against the purer quality of sugar it produced to protect the British refiners (Adamson 1972: 169–73; Beachey 1957: 68). The Usine Ste. Madeleine which opened in Trinidad in 1872 was the first central factory in the British Caribbean (Haraksingh 1984: 135). The industry survived the crisis of 1884 by reducing labor costs but also by improving industrial efficiency. Multiple-effect evaporators, extended mill tandems and mechanical conveyers to carry the bagasse from mill to furnace became standard equipment. There were professional chemists to oversee operations on the better estates by the 1880s (Adamson 1972: 185, 190–2). By 1914, there was an extensive network of private railways to carry the cane built in conjunction with the extension of the central factory system (Haraksingh 1984: 136–7).

As elsewhere, this transformation in technology increased the optimum size of plantations and changed the capital requirements of participation in the industry. The numbers of plantations declined, but their average size increased. Only the rare resident planter could finance the changes necessary to remain competitive; the majority turned to their British agents for funds, and as debts accumulated the agents acquired the plantations. These agents in their turn amalgamated to form prominent merchant houses, a process that led in 1866 to the creation of the Colonial Company which owned large estates and factories in Trinidad and Guiana. Another series of amalgamations among West Indian merchant houses produced Booker Bros., McConnell and Co. which was to dominate the Guianan sugar industry until nationalization after independence in 1966. Thus ownership of the industry passed from residents to aliens, from individuals to incorporated companies. These companies vertically integrated the industry, owning the cane fields and factories, supplying machinery, transporting and marketing the sugar. The experience of Guiana provides another example of the fact that the great capital cost of the nineteenth-century technological innovations forced a change in the ownership of the industry: trans-national companies could raise the money to buy the machinery; the local planters could not and so sold their lands to the companies.

Peru is another sugar producer that belongs to this category of new entrant on the international scene, although the fit is a little uncomfortable. By 1800, the Peruvian sugar industry was already some two and a half centuries old, but during the nineteenth century the industry expanded greatly in size and made the transition from supplying a protected regional market to competing in the international market. In the early 1860s, Peru exported a few hundred tons of sugar annually, but by 1914 trade had increased to 176,670 tons, of which Chile took 41% and Britain 27% (Albert 1976: 4a–25a). While this level of export was not large compared to the major producers, nevertheless, given the very competitive nature of the international sugar market, it represented a considerable achievement that ranked Peru above some of the traditional Caribbean exporters. The Peruvian industry, too, attracted foreign capital and became, in part at least, foreign owned.

Among the several factors behind this success, improvements in transportation as well as changes in duties and tariffs were of key importance. During the colonial period, and even up to the mid-years of the nineteenth century, the cost of transportation effectively limited sales of Peruvian sugar to Peru and Chile, and likewise transportation costs also excluded other producers from this market. The small size and protected nature of the market ensured that the Peruvian industry remained modest in output and backward in technology. Indeed, the only competition the Peruvian industry experienced was on the far side of the Andes where sugar from northeast Brazil encroached on Spanish American prerogatives in present-day Argentina. During the second half of the nineteenth century, improvements in shipping and, later, the opening of the Panama Canal lowered the cost of transport to Europe and eastern North America with the result that the Peruvian sugar industry began to make sales there. In England, these sales were aided by the abolition in 1874 of import duties on all types of sugar. But the lower transport costs also broke the protective barrier around Peru's Pacific coast market so that in the 1880s highly subsidized German beet sugar managed to undercut Peruvian sales in Chile. In 1893, however, the Chilean government began to discriminate against beet sugar when to protect a new domestic refining industry it revised tariff policy to favor the importing of the type of unrefined sugar that Peru could provide. The Brussels Convention and the resulting decline in the export of beet sugar had the double benefit for Peru of helping its exports to England as well as further curtailing German competition in Chile. On the other hand, Peru found sales to the United States more difficult to achieve after the

United States had acquired Puerto Rico in 1899 and granted Cuba tariff concessions on its sugar in 1903. In sum, however, Peru gained rather more than it lost through these changes in tariffs and duties.

Of crucial importance also was the ability of the sugar planters to find capital to finance the expansion and modernization of the industry. An early infusion of funds had its origins in the guano boom. The revenues the government derived from the export to Europe of this fertilizer that had accumulated over the centuries on the off-shore islands permitted it to compensate owners of slaves on abolition in 1854, and to pay planters premiums for every non-slave laborer they brought into the country. One president, José Echenique (1851–4), even decided to repay in cash the government's internal debt, to the great benefit, of course, of his friends and political allies. It is estimated that two-thirds of the money made available in these ways was invested in the sugar industry (Gonzales 1985: 21–3; Macera 1974: lxv). Another source of capital was the banks that developed rapidly during the 1860s, thanks to the flow of funds into the country from the export of guano, and later of nitrates. The first Peruvian mortgage banks, the Banco de Crédito Hipotecario and the Banco Territorial Hipotecario, founded respectively in 1866 and 1870, were particularly prominent in making loans to the sugar industry. Import–export houses, many of them British or American, attracted originally to Peru by the guano trade, also made loans to sugar planters, bought their sugar and marketed it abroad. This apparently smooth progress was interrupted in the 1870s with the onset of the world depression of 1873, and by the exhaustion of the guano deposits. The War of the Pacific (1879–83) dealt two more blows to the Peruvian economy: the invading Chilean armies destroyed railways and sugar mills, and in the peace settlement, Peru lost to Chile much of its nitrate-rich territory. Sugar exports dropped to about half what they had been before the War (Fig. 7.2) and only began to increase again towards the end of the century (Gonzales 1985: 33–5).

The sugar industry of Peru was indeed an attractive investment for foreign and domestic capital, and not only because new means of transport had increased the range of export possibilities. In the post-slavery era, it was able to secure the advantage of cheap labor. For a time, the Chinese were crucial to the work force, and even into the 1880s, after the end of the "coolie trade," many an estate depended on still-indentured and time-expired Chinese labor. Planters continued to pressure the government to permit the renewal of imported indentured labor, and the government finally agreed to Japanese immigration in 1898, but in effect, since the 1890s, native Peruvians had entered the plantation work force. Some were local in origin – former small-holders proletarianized by the expansion of the plantations – but most were contract laborers from the Andes who found seasonal wage labor on the coast and the agricultural routine of their mountain homes a workable if not attractive combination (Albert 1984).

By 1914, the sugar industry was one of the modern sectors of the Peruvian economy, forming enclaves of capitalist, export-oriented agriculture in the irrigated coastal valleys mostly to the north of Lima. As in other regions where the industry had overcome the difficulties of the nineteenth century, the Peruvian industry had attracted both foreign labor and foreign capital. There was British, German as well as American investment, and W. G. Grace and Co., of New York, the shipping company that was to become a diversified trans-national with a big interest in cane sugar, bought its first Peruvian sugar plantation in 1882. Yet most of the industry had remained Peruvian owned, drawing finance from Peruvian banks, and the richest planters formed part of Peru's business and political elite (Gonzales 1985: 41, 44). In reflecting on the success of the Peruvian industry in contrast to the decline in the industry of northeast Brazil, the importance of the guano boom stands out, as it provided capital and a banking and trading infrastructure from which the sugar industry benefited. Furthermore, the sugar industry was one of the few activities in which to invest money in Peru. In Brazil, the nineteenth-century booms in rubber, cocoa and coffee, and the beginnings of the industrial development of São Paulo, all offered better investment prospects than the tired, conservative plantations of the northeast: the presence of these alternatives therefore worked against the interests of the northeastern sugar industry.

Producing for the home market

Throughout the nineteenth century far more people in America were engaged in the cultivation of sugar cane for their own use or for sale in local or national markets than were employed in producing sugar for export. As sugar had become a staple in the diet of the industrial worker, so even earlier had sugar and rum become staples in the diet of Latin Americans, both rich and poor. Peasants worked small plots of sugar cane, and on many a large remote hacienda a few fields were devoted to the crop. The roughly refined rum, the product of primitive stills, and known as *aguardente* (*aguardente* in Portuguese) as well as by numerous colloquialisms such as the Brazilian *cachaça* and *pinga*, was without rival in Latin America as a cheap intoxicant except in Mexico where mescal was popular. Crystalline muscovado sugar was made according to techniques that had changed little if at all since early colonial days. It was formed into brown cakes that were, and indeed still are, a familiar sight on market stalls in country districts. This crude sugar, too, goes by various names of which perhaps the most familiar are the Brazilian *rapadura* and Colombian *panela*. It provides above all cheap caloric energy whether chewed as a candy bar, dissolved in water and drunk or eaten, as in northeast Brazil, mixed with manioc flour. Cooks in the great plantation houses evolved many recipes based on the abundance of *rapadura* for rich desserts as well as for various fruit preserves that have a close historical kinship with British breakfast marmalade. The popular taste for this archaically made sugar persisted even after fully refined centrifugal sugar became readily available, and thousands of small mills continued in production to satisfy the demand.

A second market, one for pure white refined sugar, began to develop in Latin America during the nineteenth century, at first in the large cities where the Latin American national elites were congregated, who at this time were adopting some French and English tastes in food as well as in dress, art and architecture, and where also European immigrants and businessmen had come to live. Refined sugar could, of course, be imported from Europe, although such a trade had a strong coals-to-Newcastle aspect to it as well as being a drain on foreign exchange. Alternatively, Latin American countries could build their own refineries, supply them with their own sugar and, if necessary, import the fuel they required. This became an increasingly practical proposition as the building of railways made possible the cheap and efficient transport of raw sugar from inland centers of cultivation to refineries in the main cities. Gradually, over the years, the central factory system relegated the manufacturers of such sugars as *rapadura* and *panela* to the hilly terrain or other locations unwanted by the modern industry.

By the beginning of the twentieth century three important modern sugar industries had emerged that catered strictly to national urban markets: those of Morelos in Mexico and Tucumán in Argentina were transformations and expansions of industries established in colonial times; the industry in the southern United States was a post-1800 development. The history of each of these industries was profoundly marked by the political and economic life of the national societies to which they belonged.

For the Mexican sugar industry the years between 1790 and 1914 were a time of troubles that saw only one brief period of prosperity, and closed disastrously in one of the world's major revolutions. The wars for independence caused a good deal of destruction to plantation buildings and machinery, and republican Mexico's decision in 1822 to cut off trade with Spain ended the modest flow of exports (Sandoval 1951: 169–70). The political turmoil of the succeeding years militated against the recovery of the industry which consequently grew backward as the supply of local markets did not provide any stimulus to change. In Morelos, Otaheiti cane only replaced Creole during the mid-years of the century (Melville 1979: 33), a date which gives Mexican planters a strong claim to have been among the last to have abandoned Creole cane. A significant mid-century development that later helped the aggrandizement of sugar plantations was the introduction of new laws governing landownership by liberal politicians with firm views on the virtues of private property. These laws attacked not only the church which by the "Lerdo Law" of 1856 was forced to sell its vast landholdings but also other corporate and communal owners of land such as Indian villages. The construction of railways was another break with the past, reducing transportation costs and giving some promise of the creation of a unified national market. The sugar planters had maintained sufficient influence with government to ensure that Mexico maintained a high protective tariff on imported sugar, and so when Porfirio Díaz, who was to be president for thirty-five years, came to power in 1876 promising political stability and policies favorable to economic development, the modernization of the sugar industry at last moved ahead.

In Morelos, the process was rapid, and dramatic in its denouement. Throughout the colonial period, Morelos had been the most important sugar-producing region of Mexico thanks in good measure to its proximity to the largest market in the country, Mexico City, and even at the beginning of the twentieth century it still accounted for one third of the country's production. Morelos made only 9,900 tons of sugar in 1870; thirty years later in the harvest of 1908–9 the total was 52,230 (Melville 1979: 34), but because this increase in production was achieved at the expense of the interests of the rural workers it contributed in no small measure to the outbreak of revolution in 1910. The railways that had formed a national market for the Morelos industry threw out of work thousands of muleteers and carters; even more of these workers lost their jobs when, after 1900, the planters introduced narrow-gauge railways to move cane from the fields to the factories (Warman 1984: 168–9, 172). The economic position of the rural population further deteriorated when Díaz' Minister of Public Works sold to the planters almost all the remaining public land in Morelos, and ensured that the planters, in their disputes with villages over the ownership of land, received favorable decisions. Gradually the villagers lost their lands (Womack 1969: 44).

The dispossessed villagers along with the out-of-work muleteers and carters became sharecroppers on the poorest fields of the large estates. Others hired themselves out as day laborers, working a twelve-hour day for a wage that over the years increased rather more slowly than the price of consumer goods, and with no guarantee of work except during the harvest. Many fell into debt. Numbers of workers moved onto the plantations with their families to become permanent dependent workers, receiving a house, minimal medical attention, occasional bonuses and a small regular wage the year round. These resident workers were known as *realeños*, and as a class they had existed since colonial times, but the modernization of the industry reduced their status. Formerly, they had supervised day laborers and now they were the laborers; formerly, too, they had occupied the more skilled positions in the milling and manufacture of sugar, now, without the education to manage the machinery of the central factories, they saw foreign technicians take over. Other losers in this transition were the owners of small estates who could not raise the capital to establish central factories and sold their properties to more successful neighbors. The successful plantations became very large indeed, with permanent resident populations ranging from 250 to 3,000, and by the beginning of the twentieth century there were only seventeen major plantations in Morelos. Not all the land on these plantations was in sugar, but control of this vast acreage was essential to the planters as a means of securing a cheap, dependent labor force.

In the last years of the nineteenth century, the world seemed to have become ordered to favor the Morelos sugar planters, but the capital costs of modernization posed a strain on even the most successful, and after 1900 the financial pressure became even more intense. Other long-standing regions of sugar production, Puebla and Michoacán, continued to pour sugar onto the Mexican market, and they were joined by new producers in Sinaloa and by American-financed enterprises in coastal Veracruz. The domestic market became glutted, and Mexico began to participate in the highly competitive international market. To reduce costs, planters had to make their operations more efficient, and one step in this direction was to make full use of their new large mills and factories. This required them to put yet more land in cane, increase their intake of water for irrigation, and so annex even more of the physical and human resources of Morelos. In 1908, the government eased the situation of the planters by doubling its tariff on imported sugar, thereby permitting an increase in the domestic price, which in turn prompted a further expansion in sugar production. This high degree of social injustice could not be maintained, and during the trauma of the Mexican Revolution and the land reform that followed the Morelos sugar industry was swept away, the mills and mansions destroyed, the fields returned to villagers (Warman 1984; Womack 1969: 37–66, 373–4, 376).

In the context of Latin American history, late nineteenth-century Morelos provides perhaps an extreme example of what happened when improvements in transportation opened formerly isolated regions to capitalist exploitation: the commercialization of agriculture took place at the expense of the interests of the local population. There was, indeed, progress of a kind, for the small minority, but also poverty and the destruction of a way of life for the majority (Burns 1980). In the context of the history of the sugar cane industry, Morelos provides an instance of the failure of planters to resolve the labor problem; as in the case of St. Domingue more than a century before, the exploited work force at the opportune moment rose up and destroyed the plantations.

Argentina provides a classic case of the roles of improved transportation and nationalist policies in the development of a sugar industry. The regions of Argentina with a climate warm enough for the cultivation of sugar cane and with the necessary water supply lie in the far north and northwest of the country towards the borders of Bolivia and Paraguay, 1,000 km. and more from the great market of Buenos Aires. Sugar cane was first introduced to this remote part of the Spanish colonial world shortly after the conquest, but the expense of transportation to outside centers of population ensured that production was very small scale and for local consumption only. Nothing occurred to change this situation until the third quarter of the nineteenth century. Then the civil wars and secessionist movements that had characterized Argentina since independence ceased with centralists in the ascendance and Buenos Aires as the undisputed dominant city. The destruction of the remaining Indian tribes in the center and south of the country between 1879 and 1883 in the war known as the Conquest of the Desert ended Indian threats to the safety of transportation and settlement, and the coming of the railway made it possible to think realistically in terms of creating a single, national market. In other words, the conditions now existed to promote the development of a sugar industry.

The historical geography of sugar production from the 1870s onwards is intertwined with the political history of Argentina. Even the rise of the province of Tucumán to the dominant position in the industry owes much to the geopolitical strategies of the elite – the so-called Generation of 1880 – that ruled Argentina between 1874 and 1916. It followed policies of national integration and modernization through the building of railways, the creation of a national banking system and promotion of the agricultural potential of the interior. The northwest had strong links with Bolivia and Peru which since colonial times it had supplied with agricultural products. To reduce the risk of secessionist movements, the national government wished to tie the northwest and its regional capital, the city of Tucumán, closely to Buenos Aires and the national economy, and a sugar industry seemed one means to this end. A sine qua non for the success of such a policy was transportation, and accordingly, in 1870, the government authorized the prolongation of the railway from Cordóba to Tucumán, where it arrived in 1876. In this same year, further assistance came to the industry from Congress which passed tariffs against foreign sugar and exempted machinery for manufacturing sugar from the tariffs on industrial goods. From the 1880s, again at the instigation of the government, the railroads charged low rates for transporting sugar. These various measures had the desired effect of making Tucumán sugar competitive with foreign in Buenos Aires, and production expanded from 1,000 tons in 1870 to 270,504 in 1914. The political representatives of Tucumán in the national government became able lobbyists for the sugar industry (Guy 1980: 3–4; 1984: 151).

In contrast to so much of the economic development in late nineteenth-century Argentina, the Tucumán sugar industry was in good measure Argentinian in its entrepreneurs, finance and labor. The regional landowners did not themselves have the resources to bring into cultivation sufficient land and to construct the central factories necessary to meet the national demand for sugar, and they were joined by investors from Cordóba and Buenos Aires, some of whom moved to Tucumán. The most visible foreign interest was French. During the first half of the century, a few French emigré families had settled in Tucumán where they had become involved with sugar production after the 1870s, and this interest was reinforced through the commercial acumen of Baron Portalis who was the energetic representative in Buenos Aires of two French companies important for the sugar industry: Fives-Lille, the manufacturers of equipment for sugar mills, and Decauville, the manufacturers of narrow-gauge railways. With loans from national banks as well as the assistance given to some planters by Portalis, the new Tucumán industry took on a modern aspect (Guy 1980: 48–52).

The emergence of agribusiness in a formerly sparsely populated region brought with it a shortage of labor. Immigrants were flocking into Argentina to help make the Pampas one of the most productive agricultural regions of the world, but they could not be tempted to move north to work in the subtropical cane fields. The authorities in Tucumán adopted various strategies that included draconian legislation to compel so-called vagrants and supposedly unemployed people to work in the cane fields as well as the use of the forced labor of Indians captured during the Conquest of the Desert. Neither source sufficed (Guy 1980: 34–6). The solution to the problem lay in the rural population of the countryside of northern Argentina, poor people of mixed Spanish and Indian descent. The slack period on their small farms – the dry season – coincided with the harvest on the sugar estates, and there began an annual migration of seasonal workers that by 1895 accounted for the greater part of the 60,000 laborers employed in the Tucumán harvest (Kirchner 1980: 15–18). Non-Argentinian labor appeared only in the far north, in the provinces of Salta and Jujuy, where estates employed Indian laborers from Bolivia (Whiteford 1981: 32–3). By the early 1890s, the new Argentinian sugar industry would seem with strong government support to have overcome relatively easily the problems of finance, labor and market that all cane-growing regions faced, but this very success soon brought difficulties of its own, and led the government onto the tortuous path of managing a surplus of sugar and contending with various interest groups that now surrounded the production of sugar.

The aim of the Tucumán producers was to preserve the Argentinian market for their sugar and, after the large harvests of the mid-1890s when they began to produce more sugar than Argentinians could consume, to ensure government assistance in the export of sugar. Consumers and their politicians in Buenos Aires objected to the high cost of domestic sugar and favored the reduction of tariffs on cheap foreign supplies. The situation was further complicated by the construction of a refinery during the late 1880s that was to make the country self-sufficient in refined sugar. The refinery had a guarantee from the government of a minimum level of profit, and this required agreement on a price for raw Tucumán sugar that permitted the refinery to make its profit and still sell the refined sugar at a lower price than the imported. The refinery, in fact, managed matters rather well, driving foreign refined sugar out of the Argentinian market until 1907 when consumer lobbies began to achieve some success in their arguments for a reduction in tariffs. The campaign to export sugar faced a problem of finding a market. There was little prospect of sales to neighbors: Brazil was self-sufficient in sugar and Peru supplied Chile. Exports supported by a bounty could not be expected to enter Europe after the Brussels Convention of 1902, and Argentina, although not signatory, discontinued sugar bounties. Complex negotiations with the United States led to a concession of lower tariffs on Argentinian sugar provided it had not been subsidized through bounties. The Tucumán lobby still had sufficient clout to persuade a government intent, in 1912, to lower tariffs on imported sugar to do so gradually over a period of years rather than at once (Guy 1984: 147–62).

Effective lobbying on the part of producers, the dynamics of the internal politics of Argentina and a nationalistic desire to be self-supporting in an important product gave the country a modern but high-cost and protected sugar industry. This was indeed an achievement of a sort, but it left the country with two problems: disposal of surplus sugar on the intensively competitive international market through diplomacy or cajoling in one way or another and the reconciliation of the increasingly entrenched vested interests of a remote and poor part of the country in a high-cost sugar industry with those of the consumers of sugar in the national capital and other cities.

During the course of the nineteenth century, the United States of America, too, acquired a large and uncompetitive sugar cane industry that also has survived protected by tariffs into the present day. The French and Spanish who settled the lands along the coast of the Gulf of Mexico from Texas to Florida had not seriously interested themselves in sugar cane for two very good reasons: the cool winters made the Gulf coast a less than optimum environment for the industry and the Caribbean colonies were already supplying the French and Spanish markets. Travelers did indeed introduce samples of sugar cane to Louisiana at various dates during the 1700s, but only in the last decade of the century did experiments in the commercial production of sugar begin. The leading protagonist was the French owner of an indigo plantation near New Orleans who had become dissatisfied with the low price of indigo. He showed that sugar cane would mature in the Louisiana climate, but the limited scale of his activities left open the question whether or not sugar could be manufactured at a competitive price. The investigations received a boost when immigrants arrived from the wars in St. Domingue bringing with them their technical knowledge of the sugar industry while the Louisiana Purchase of 1803 opened the United States market to Gulf coast planters. These changed circumstances made all the difference, and during the first half of the nineteenth century a new cane-growing region came into being.

The extent of this region has been strictly limited by the physical geography. Both climate and terrain were and are a problem. The risk of frost was a serious hazard, and experimentation led to the planting of cane in the fall with the harvest a year later. An early frost in late November could damage unharvested cane while a late frost in March killed new cane as it appeared above ground. Severe winter frosts killed the cane in the ground. During the first half of the nineteenth century, landowners tested the climatic restrictions of the South on the sugar industry, pushing cane cultivation up the Mississippi valley and along the Atlantic coast as far as South Carolina where the politician and noted agricultural improver, Thomas Spalding, persisted unsuccessfully with attempts to produce sugar from 1806 until the 1830s (Spalding 1816; Coulter 1940). The introduction in 1817 of the Ribbon variety of cane, also known as Batavian Stripped and Black Java, proved a great boon as its thick rind made it more resistant to cold than the Creole and Otaheiti varieties, and it also matured more quickly (Sitterson 1953: 120). Nevertheless, the frequency of frost-damaged crops in the more northerly locations eventually drove home the lesson that cane was indeed a tropical plant and could not be grown with reasonable expectations of success outside southern Louisiana, the adjacent coastlands of Texas, and Florida. Even within this area, the terrain imposed further restrictions on the cultivation of cane: much of southern Louisiana in the nineteenth century was badly drained if not swamp; the Florida everglades – site of a modern sugar industry – were also still undrained and the sandy soils of the northern part of the state did not produce good crops; in Texas, only the alluvial bottom lands along the lower courses of rivers offered good prospects. The best lands for sugar cane cultivation were the levees along the main stream of the Mississippi and its distributaries, the bayous. Here, on these gentle slopes, the Louisiana sugar industry took hold.

This riverine location, combined with the physical characteristics of the levees, greatly influenced the spatial organization of production and created one of the more distinctive plantation landscapes of the nineteenth-century sugar industry. As transportation was by water, the land was divided into long, narrow lots to give as many owners as possible a narrow frontage on the river. The mansions and mills stood on the height of the levees, well above flood level, near the wharves, and where a breeze was more likely to assuage the sultry heat of the evenings while the cane fields reached down the well-drained soils towards the backswamp. On the larger levees, there was a second range of lots which were clearly far less desirable than those abutting on navigable water. The result was a very linear settlement pattern that persisted into the present century, when an expanding road network, flood control and the drainage of the swamps led to important modifications. The agricultural advantages of the levees were offset by the risk of flood. Planters worked to reinforce and raise the natural levees, but water often overflowed, damaging fields and buildings and in a bad flood even obliterating entire plantations. Care of his own levees did not necessarily spare a planter from damage because his lower fields could be flooded when a break in the levee upstream raised the level of water in the backswamp. The New Orleans merchants knew the quantities of sugar to be expected from each distributary or bayou and the risk of flood notwithstanding, the benefits of locations on the levees over other sites in the South resulted in Louisiana accounting for 95% of Southern sugar production in the years before the Civil War (Hilliard 1979; Sitterson 1953: 13–23).

The growth of the Louisiana industry depended on the ever increasing size of the United States market for sugar, on tariff protection for sugar as well as on the state of the cotton trade, its competitor for land, labor and capital. When an increase in protection coincided with a downturn in cotton prices there was a move of landowners into sugar. This conjunction occurred twice before the Civil War with quite dramatic results: in the 1820s, when the numbers of sugar plantations increased from 193 in 1824 to 691 in 1830, and again between 1845 and 1849. In 1849, there were 1,536 sugar plantations in Louisiana, the antebellum peak in numbers although not in sugar production (Fig. 7.2), which continued to increase erratically until the outbreak of hostilities in 1861 (Sitterson 1953: 23–30). At this crucial point in Southern history, the sugar industry was still unreformed: 98% of the plantations were worked by slaves, with planters owning on average seventy-five slaves except along the Mississippi where the average approached a hundred, and most planters milled and manufactured their own sugar in the traditional way although a good many had moved to steam power (Schmitz 1979).

Change would inevitably have come to the Southern industry even without the Civil War because of the various pressures of abolitionism, commercial competition, technological and scientific advance, and progress in transportation that were working on the sugar cane industry world-wide, but in all likelihood the process would have been a gradual one, marked by the selective adoption of innovations as finances, opportunity and need suggested, and without disruption in production. The War, however, dealt the old system a devastating blow through the physical destruction of military defeat and the consequent freeing of the slaves without any financial compensation to planters or even the concession of some sort of transition period for adjusting to new labor conditions. By 1865, the Southern sugar industry lay in ruins with its traditional labor force gone, its mills and equipment destroyed, and with much of the cane land under water, thanks to the neglect of the levees. During the last year of the War, only 175 plantations were still able to produce sugar. The decline in the value of capital investment in the industry from an estimated $194,000,000 to perhaps $25,000,000 or $30,000,000 indicates the extent of the financial catastrophe the planters had suffered (Sitterson 1953: 226–7). The planters in good conservative fashion at first tried to reestablish as best they could what they had known in antebellum times, and it took some years for them to realize that this was neither possible nor indeed appropriate. Attempts at continuity then gave way to the acceptance of change.

Yet, this said, the developments in the Southern sugar industry in the decades leading up to the First World War had much in common with developments elsewhere in the industry during these years. The response to abolition has a very familiar ring. Some of the ex-slaves left the plantations and those remaining in the vicinity appeared reluctant to work in the cane fields. The planters for their part considered the former slaves poor and unreliable workers and looked to coerce them into the fields through legislation that would limit their freedom of choice in the job market. They sought also to keep wages down by increasing the supply of laborers through immigration from Europe and Asia. Neither of these strategies produced satisfactory results for the planters, and they turned to what proved to be not very successful experiments with various types of tenancy before accepting the necessity of paying the competitive wages that in fact secured the return of a reliable force of black field hands (Schmitz 1979: 273–4; Sitterson 1953: 221, 231–51). Planters were conservative, too, in their attempts to remain both growers and millers of cane in the antebellum tradition, but this organizational structure gave way under the imperatives of technological advance during the 1880s and the central factory system took over. Corporate ownership gradually replaced individual ownership of plantations and factories, although the stock of a given corporation was in most instances held by the members of a single family rather than by institutions or by the general public. Nevertheless, the process of consolidation had gone so far that by 1910 it could seriously be suggested that all the sugar factories of Louisiana be brought into one large corporation (Sitterson 1953: 259–63). The search for agricultural improvement, too, was very much a part of the Southern scene, and the attempts to reduce costs through improved yields of cane and the introduction of labor-saving techniques became more organized after the founding by the Louisiana Scientific and Agricultural Association of a Sugar Experimental Station in 1885. One factor that influenced the speed with which this rebuilding and modernization took place was the availability of capital. In the immediate postwar years, with the South's financial institutions in disarray, capital was in very short supply. The slow return of capital may well have been a result of the fact that the Southern industry seemed to potential investors always to have an uncertain future because of the threat to its prosperity posed by competition from other cane sugar producers and, after 1890, from domestic beet producers as well. Certainly, towards the end of the nineteenth century and in the early years of the twentieth, Northern U.S. and even Canadian banks and entrepreneurs, to judge from the flow of funds, regarded the sugar industries of Cuba, Puerto Rico and the Dominican Republic as good prospects for their loans (Quigley 1985). Indeed, the difficulty of raising capital in Louisiana combined with the greater capital requirements of sugar production over cotton may perhaps explain why cotton crops of the South recovered to prewar levels by 1879 while sugar crops did not do so until 1894 (Schmitz 1979: 271).

The degree of protection the Southern industry received changed with the policies of the United States government and lobbying Washington became a constant preoccupation of planter organizations. They built a case for their industry on the predictable grounds of the numbers of workers it employed directly and indirectly, the business it gave to manufacturers of machinery in other parts of the country, the traffic it generated for the railroads and so forth. However, in the late nineteenth century the Southern pressure group was not the only sugar interest the United States government had to deal with: there were the investors in the United States' new tropical colonies of Puerto Rico, Hawaii and the Philippines, the owners of plantations and railroads in Cuba and the Dominican Republic, beet farmers in the Middle West and West, and also, of course, there were the consumers who wanted cheap sugar. In 1875, Congress admitted Hawaiian sugar free of duty with the result that the Hawaiian industry began to expand rapidly. In 1900, it reduced the duty on Puerto Rican sugar by 85%, and removed it altogether in 1901. Philippino sugar, too, benefited: in 1902, Congress reduced the duty by 25%, in 1909, permitted 300,000 tons to enter duty free a year, and in 1913, removed all duties and controls. A strong lobby with large investments in Cuba effectively eased the way for Cuban exports to the United States. In the meantime, the domestic beet industry grew substantially. The result of this competition for the Southern industry was a decline in its share of the expanding United States market from a little over 12% during the 1890s to about 5% in 1916 and only 1.8% in 1930, but, nevertheless, the protection it received was sufficient to permit a slow absolute increase in production until just before 1914. The peak of production came in 1904, with nearly 400,000 tons, but thereafter a slow decline soon became very marked and lasted for twenty years. Bad weather, disease in the cane, poor agricultural practices, high labor costs and low sugar prices were all contributory factors, and without a continuing measure of protection the industry might not have survived at all (Schmitz 1979: 280–5; Sitterson 1953: 324–42).

There were other regions of Latin America where central factories were replacing the traditional mills during the late nineteenth century. Entrepreneurs, both native and foreign, were willing to invest in new factories wherever markets seemed to be large enough to justify the risk, or exporting the sugar appeared possible. The growth of the city of São Paulo, for instance, was in good measure responsible for the revival of the industry. Sugar had been cultivated on a very minor scale in what is now the state of São Paulo since the beginning of European settlement, and in the late eighteenth century and the first half of the nineteenth it became an industry of regional importance, supplying principally the city of Rio de Janeiro but also exporting to Montevideo, Buenos Aires and via Rio to Europe. The quality of the sugar was poor, commanding low prices abroad, and when coffee became a more attractive investment, landowners began to cultivate this new crop instead (Petrone 1968). In the late nineteenth century, sugar production revived on the basis of central factories and wage labor, and before 1914 already accounted for 8% of the total Brazilian sugar production (Carli 1943). This percentage has continued to increase, and São Paulo has become the most important cane-growing region in Brazil. The modern sugar industry of the Cauca valley of Colombia dates from just after the opening of the Panama Canal in 1903 which together with the construction of a railway from Cali, the main town of the valley, to the Pacific coast at Bueneventura made exports possible. The Canal, by giving easy access to Europe and the eastern seaboard of the United States, also encouraged export-orientated sugar industries along the west coast of Central America. The growth of population in California has since provided another large market.

The year 1914 is a time to take stock of the American sugar industry when a long period of transition was largely over. The industry had indeed survived as an activity of major economic importance despite the problems inherent in overcoming the crises of abolition, a technological revolution and competition from sugar beet, all at a time of generally declining prices. Governments had indeed shown themselves to be ready to intervene to protect the sugar cane producers through protecting home markets, with financial assistance for modernization and by facilitating the flow of immigrants to provide cheap labor. The industry's prospects, moreover, were reasonably good. Mankind's taste for sugar was unabated, and as the population of the world increased, so the demand for sugar might be expected to continue. The Brussels Convention had put in place an internationally agreed mechanism that was expected to prevent prices from falling again to disastrous levels. There was every reason to suppose that this nexus of demand and support would continue, and that sugar cane would be able to hold in the future a good share of the market for sweeteners. One measure of the success with which the cane industry had overcome the difficulties of the nineteenth century was the increase in production, from approximately 300,000 tons in 1790 to nearly 5,000,000 in 1914.

However, the expansion and modernization of the industry had costs for American society that are difficult to express in financial terms. Slavery, it is true, had gone in the legal sense of planters owning their laborers and this was an enormous plus, but few of those who toiled in the cane fields in 1914 had achieved any real economic freedom. Indenture arrangements, although declining in importance and soon to be done away with, still bound many workers to the industry, while debt peonage and other more subtle forms of coercion still weighed on the work force. More and more of the rural people of Latin America, the "folk" as Burns (1980) calls them – the remnants of Indian societies and peasants of mixed Indian, African and European descent – had been entrapped in the ever widening net of the plantation economy and forced to work in the cause of sugar as railways extended into the interior and planters bought or seized more land. Sugar was "the only tropical crop to experience a scientific revolution before W.W.1" (Lewis 1970: 19) and although the central factory system had been essential for the survival of the industry in its struggle with beet sugar, it contributed to that change in the ownership of the industry from the local planters to large corporations, often foreign-financed. True, the exports brought revenues to national exchequers, but in several countries they also made national income depend on the market price of this one crop. The survival of the American sugar cane industry in this manner is a reflection of the fact that the political struggles of nineteenth-century Latin America produced neither true individual freedom for the rural workers nor economic freedom for the countries.

PART 2

The sugar industry in the East

8
Asia: *c.* 1750–1914

During the centuries between the establishment of the sugar industry around the Mediterranean and the beginnings of European imperialism in Asia, the sugar industries in the East and West existed independently of each other. Indeed, after that initial diffusion of sugar cane and sugar technology from India to the Levant, no record exists of any significant exchange between East and West of agricultural or manufacturing information on the production of sugar until the vertical-roller mill reached Peru from China at the beginning of the seventeenth century. This mill proved to be Asia's last technological gift to the western sugar industry. Perhaps the best explanation for this lack of contact over so long a period of time lies in the fact that the two industries operated in contrasting cultural and economic environments each of which had its own imperatives for sugar production; consequently, the innovations in the capitalist and imperial western industry that led to an ever increasing scale of mass production had little relevance to the eastern industry geared to service of local economies. If in Asia there had been any increase in per capita consumption of sugar over the centuries, it was certainly very modest and not at all comparable to the upward spiral in the West. For whatever reason, the Asians had not become "hooked" on sugar cane and so the demand did not exist to support an expanding industry. Capital investment in Asian sugar production was minimal, cultivation a peasant activity, the technology small scale and slow to change. In all of southern Asia, only a few regions produced sufficient sugar to support an export trade. All this was to change as Asia fell victim to European imperialism and was drawn into the economic system centered on the North Atlantic: its sugar industry was westernized, the economic and social geography of the cane-growing regions gradually transformed. Under the British, production increased in India, particularly in Bengal; the Dutch turned Java into a major exporter, Taiwan became a sugar colony of the Japanese whose taste for sugar grew as they industrialized and became more prosperous, and in the Philippines a backward industry began to expand and modernize as opportunities for export arose but it only really flourished after the Americans took control of the islands from Spain. In China, although not formally part of any colonial empire, sugar production too began to respond during the nineteenth century to the pressures of the outside world. By 1914, the geography of the Asian tropics had come to owe a good deal to the world's ever increasing demand for sweet food and drink.

India

The Indian sugar industry *c.* 1750, when the British began to take over the administration of parts of the sub-continent, and even later was as far as can be established very little changed from much earlier centuries, and in this respect it was no different from the rest of Indian agriculture which was technologically stagnant compared to Europe, China and Tokugawa Japan (Raychaudhuri 1983: 15–16). Inevitably, the British commented on the backwardness and inefficiency of sugar production. Francis Buchanan, a traveler in India who admitted to a "very slight knowledge of Jamaica," was struck by the "wretched state of mechanics" (1807: vol. 1, 342): mortar and pestles and simple two-roller mills of very limited capacity were in common use (Plates 9 and 10) while the equipment in the boiling houses was "extremely crude" (1807: vol. 2, 276). Moreover, the local varieties of cane were of poor quality and yielded badly. Sugar cane, nevertheless, was widely cultivated in most parts of India but only as one of many crops in a village. The principal product was still *gur* (referred to by some authors as *jaggery*) for local consumption. In the towns, refiners produced a coarse sugar known as *khandsari* as well as some superior qualities for the luxury trade. Delhi had important sugar refineries (Naqui 1986: 171), maintained no doubt by the demands of the Moghul court; Ahmedabad was known for the manufacture of loaf sugar (Gupta 1970: 112), and by the late eighteenth century some western influence was apparently beginning to make a mark, for Buchanan (1807: vol. 1, 340) noted that around Seringapatam in southern India some clayed sugar, "fine and white," was being made, the "art" having been introduced by the sultan (unidentified but perhaps a reference to Tipu Sultan, of Mysore, who died at the siege of Seringapatam in 1799). The main regions of sugar production, however, had long been in the north, in Punjab, the Ganges valley and Bengal, and under the stimulus of the new market forces the acreage of cane had increased during the seventeenth and early eighteenth centuries. Some small part of this northern production entered the export trade via the ports of Bengal and, on the other side of the sub-continent, of Surat and Cambay, finding markets in Persia and the Middle East. This longstanding trade lasted until the mid-eighteenth century when political turmoil

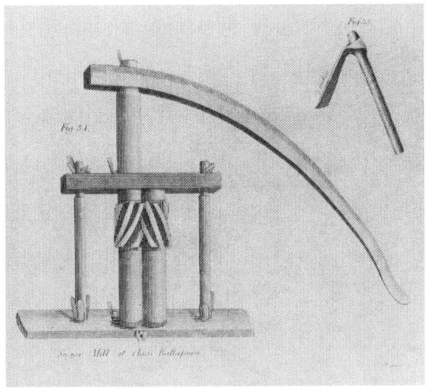

Plate 9 Two-roller mill, South India, early nineteenth century
Source: Buchanan 1807: vol. 1, 341.

in the Middle East brought it to an end (Chaudhuri 1983: 321; Chaudhuri 1985: 193–5; Gupta 1970: 111–12; Raychaudhuri 1982: 273–4; Raychaudhuri 1983: 15).

The East India Company attempted to revive the commercial production and export of sugar, concentrating its efforts at first in Bengal and aiming at markets in Europe. Developments at the end of the eighteenth century suggested that the time was opportune for this initiative. The passage in England in 1774 of the Commutation Act, which greatly reduced the duty on tea, had led to an increase in consumption not only of tea but also of sugar, and, after 1791, prices for sugar rose in response to the slave revolt in St. Domingue. In Bengal, Company officials set about making cultivators aware of the new market opportunities while at the same time they tried to prevent landowners from exploiting the situation by raising rents on sugar lands. The Company also encouraged Europeans to invest in the industry in the hope that they would introduce West Indian technology.

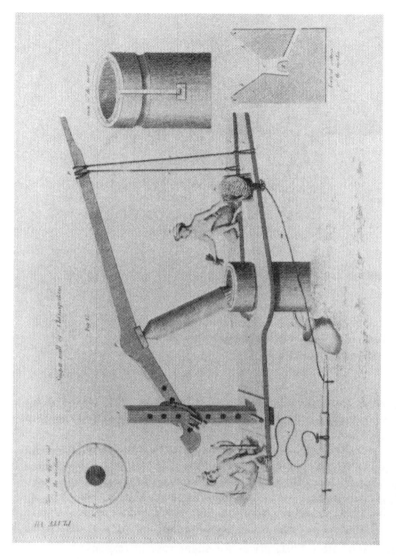

Plate 10 Mortar and pestle, South India, early nineteenth century
Source: Buchanan 1807: vol. 1, 159.

The policy initially had some success with exports reaching 4,000 tons in 1804, but the Napoleonic Wars cut off the continental markets and penetration of the British market was made extremely difficult by duties that still favored West Indian sugar. After 1811, with exports down to 120 tons and the European investors discouraged, the Company ceased to ship sugar from Bengal to Europe. The end of hostilities reopened the prospect of exports to continental Europe but in Britain the power of the West Indian lobby continued to frustrate for some time the Company's attempts to reduce the duty on Bengal sugar. In 1824, the duty was actually increased by 100%. Eventually, Bengal began to benefit from the anti-slavery movement for as this became more powerful in Britain so the influence of the West Indian lobby waned. In 1830, the duties on Bengal sugar were halved and in 1835–6 reduced to parity with those on West Indian sugar (Chaudhuri 1983: 321–3).

The trade statistics reflect the opening of the British market to Bengali sugar. In the decade following 1835–6, exports rose from 7,300 tons to 68,000, quantities that represented respectively 8.7% and 37% of the total value of the exports passing through the port of Calcutta (Chaudhuri 1983: 323). For British India as a whole, sugar amounted to only 3% of the value of all exports in 1814–15 but its share had risen to 10% by mid-century (Chaudhuri 1971: 26). While India had perhaps not become in any real sense a sugar colony, it was, nevertheless, a significant presence in the international sugar trade and the sugar industry had become an important sector of its economy. Yet India was not able to maintain this success in the international market. During the second half of the nineteenth century, exports ceased and India even became an importer of centrifugal sugar. A part of the explanation for this change is the familiar one common to all producers of cane sugar: competition from beet. A second factor adversely affecting producers in India as elsewhere in the British empire was Britain's adoption of free trade. A third factor must surely have been the nature of the Indian industry itself. The sugar industry in India was throughout the nineteenth century largely a domestic industry and seemingly resistant to change. As the population grew, the internal market for sugar did increase. The government's insistence on the cash payment of taxes certainly encouraged the peasants to put at least some of their land in commercial crops such as sugar. The extension of irrigation systems improved yields in the Punjab and elsewhere; better roads and later the railways helped bring the sugar into the towns. There were, on the other hand, some negative features that varied in their importance from one part of India to another according to environmental and social conditions. The labor demands of sugar cane cultivation were high compared to other crops and there was a limit to the amount of labor a peasant family could provide. For those with only small farms from which to feed their families, sugar cane represented a heavy commitment in that it occupied the land for more than a year. Also, it needed manure which was usually in short supply and on which there were many other demands. In effect, many landowners with limited resources had to borrow money to finance a crop of sugar cane. Profitability was a consideration of basic importance: unlike a West Indian planter, often an Indian peasant had little if any capital investment in mills or machinery for sugar production and so could turn easily to wheat, cotton, indigo or some other crop if it gave a better return. In these circumstances, many manufacturers of *khandsari* were also moneylenders securing the raw material of their business from dependent indebted peasants (Banerjee 1982: 63–6; Amin 1984: 36–8, 41–61).

An important question about this domestic industry is why it remained by western standards so technologically backward. In much of the countryside throughout the nineteenth century, the standard press was still a form of mortar and pestle for which the cane had to be chopped into small pieces. The model shown in Plate 11, known as a *kolhu*, was turned by two bullocks (Amin 1984: 54). In the 1870s, Messrs. Thompson and Mylne, sugar planters in Bihar, began manufacturing a small, portable, bullock-powered mill consisting of two iron rollers vertically mounted in a wooden frame supported on four legs. This design soon replaced many *kolhus*. During the 1870s, a hand-driven centrifuge also came into use (Amin 1984: 124–7). Despite the knowledge of government officials of West Indian technology and even the arrival in India at mid-century of some West Indian planters disillusioned by post-emancipation Caribbean life (Chaudhuri 1983: 323), neither the old three-roller mill nor the new horizontally mounted roller mill were adopted in India. What is the explanation for this technological conservatism? The answer probably lies in a combination of factors: in the structure of rural society, in the patterns of land tenure, in the food tastes and the low purchasing power of the rural population, in the better opportunities for capital investment in other forms of economic activity. Possibly the scale of the peasant operations was too small to justify the use of West Indian style mills, the means of transport inadequate to the task of assembling a sufficient quantity of cane at a given spot to keep them in operation. Amin (1984: 104) suggests that religious prejudice against "machine-made sugar" slowed the acceptance of innovations. The argument (p. 207 below) that technological change is difficult in societies dependent on wet rice cultivation is not applicable to all the regions of India in which sugar cane was grown. Only in the early years of the twentieth century did central factories make their appearance in India, manufacturing sugar for the urban and European populations of the country. Although by 1914 the acreage of sugar in India had greatly increased since the European arrival and production had become more commercial, the persistence of traditional techniques and qualities of sugar suggests the Indian industry

Asia: c. 1750–1914 203

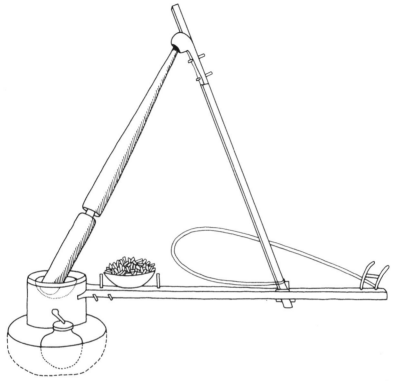

Plate 11 Mortar and pestle, United Provinces, India, early twentieth century (the Gorakpur Wooden Kolhu)
Source: Amin 1984: 54.

rather than succumbing to the forces of modernization that were producing the global sugar economy had been successful in resisting them.

China

The Chinese industry also remained outside the mainstream of international commerce until 1914 and even after. Although sugar cane was widely cultivated in southern China, except in a few localities it was a small-scale, mainly peasant activity that provided children who chewed much of the crop with a simple, cheap, sweet treat; and the manufacture of sugar was at best a minor activity. The availability of cane sugar had not in any way transformed Chinese cooking or entirely replaced other sweeteners such as maltose and honey. The Chinese continued to have

as a whole ... perhaps the least sweet tooth of any people on earth, until quite

recently, when the blandishments of soft-drink and candy hucksters ruined their resistance and, consequently, their teeth. Fruit was a light snack, not made into fancy desserts, and was liked in a fairly sour state; many Europeans who commented much on food in early accounts of modern China complain of being given green fruit.
(Anderson and Anderson 1977: 333, 344)

Indeed, as any patron of a Chinese restaurant knows, tea is taken without sugar.

There had been periods of considerable expansion in the centuries before the appearance of European traders in the China Sea, and during Sung times (A.D. 960–1279) in particular – as part of what Bray (1984: 600) calls the Sung "Green Revolution" – government encouragement had helped sugar cane become a cash crop in Fukien, Kwangtung and Szechwan. In some prefectures of these provinces sugar cane appears to have been the dominant crop. A monograph on sugar dating from the twelfth century A.D., the *Thang Shuang Phu*, reports that in one area of Szechwan no less than 40% of the peasants cultivated sugar cane (Bray 1984: 601); in Fukien, sugar cane, already prominent in coastal prefectures, began in Ming times (A.D. 1368–1644) to supplant rice as the main crop of even inland prefectures. As sugar cane and rice competed for labor, water and fertilizer, the success of sugar cane reflected an increase in the profitability during these years of the sugar industry of Fukien. Fukien sugar was traded in China, but also found markets in Southeast Asia (Rawski 1972: 48–50), a situation that lasted until the Dutch developed plantations in Java.

An important late Ming (1637) work on agricultural technology, Sung Ying-Hsing's *T'ien-Kung K'ai-Wu* (*The exploration of the works of nature*), and which Needham (1965: 171–2) has described as "China's greatest technological classic," provides useful insights into Chinese sugar production around the time of the European arrival in the Far East. Sung's report that the two provinces of Fukien and Kwangtung alone accounted for nine-tenths of all production is further evidence of the importance of the south coast to the sugar industry. Two varieties of cane were grown, one for chewing and the other for making sugar, and although he does not provide the botanical detail with which to make an identification, both certainly were varieties of *S. officinarum* L. His discussion of sugar cultivation shows a concern for frost which is indeed a risk in Szechwan and in the northern interior of Kwangtung and Fukien but not along the coast from Hainan to Fuzhou. The seed cane was buried under dry earth during the winter to protect it from the cold, and then cut into seed-pieces which were planted close together in nursery plots in the spring. Transplanting into the carefully spaced furrows of plowed fields took place when the shoots were 6 to 7 inches high. The procedure is reminiscent of planting cane in another region with risk of frost: southern Spain. The seed-pieces were at first

Asia: c. 1750–1914   205

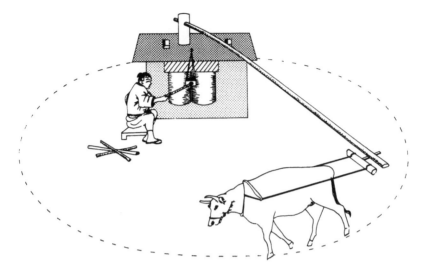

Plate 12 Two-roller mill, China
Source: Sung Ying-Hsing 1966: 127.

only lightly covered with soil, but as the shoots grew longer more soil was added around the base of the plants to encourage the development of deeply rooted tall plants, unlikely to topple. The fields were carefully manured according to the fertility of the soil and the space between the rows of cane was frequently worked, even plowed, presumably, although Sung does not give the reasons, to remove weeds and conserve soil moisture. At the beginning of the ninth month in frost-prone regions, more earth was heaped about the roots to protect them from frost and snow of the coming winter – from which it can be inferred that ratooning was part of the agricultural practice – and the cane was then harvested late in the ninth or early in the tenth month. The timing of the start of the harvest was a bit of a gamble: delay meant a longer growing period and hence a less immature cane, but it could also mean frost damage and loss of all or part of the crop. Each household that planted more than ten *mou* (1.7 acres) of cane had its own mill and boiling equipment in order to deal with its crop quickly, within ten days. This last comment is the only one Sung Ying-Hsing makes on the social organization of sugar cultivation; on the south coast where the frost-free climate permitted the cultivation of mature cane and ease of access to the sea permitted an export trade, the scale of individual operations may have been larger.

Sung describes only one design of mill, a vertical two-roller mill shown in Plate 12. This is, according to Needham (1965: 204), the first Chinese illustration of a sugar mill. Sung claimed that it worked on the same principle as a cotton gin, with the difference that the rollers in the gin are mounted horizontally and in the sugar mill vertically. Indeed, Needham (1965: 204) considers the gin and the sugar mill to be the earliest representatives of roller mills but that neither was likely to have been of Chinese origin as both cotton and sugar cane cultivation came to China from South Asia. Almost certainly, this Chinese roller mill, probably of Indian derivation, was in turn the inspiration of the three-roller mill of the colonial American industry. Sung does not explain the purpose of the frame through which cane was inserted: it may, however, have helped reduce arm-fatigue and at the same time protect the worker from trapping his hand in the rollers. It was not maintained in the American mills and seems to have dropped out of use in China as it is not shown in more recent illustrations (Hommel 1969: 113–14). Sung's mill gives the impression of slow-paced, rather leisurely activity compared to what went on in colonial America, but as each cane was pressed three times presumably such mills were worked around the clock in order to deal with even the small harvests of the Chinese peasants.

Sung describes the manufacture of three types of sugar. The first of these he refers to as "brown" sugar and would appear to have been a somewhat improved version of *gur*. Three pots were arranged over a fire with bagasse as fuel. In two of the pots, raw juice with the addition of some lime as a clarifier was boiled until it had been greatly reduced, when it was then transferred to the third pot for further reduction into the final crude sweet concentrate. Meanwhile, the first two were refilled with more raw juice, and so the process continued. The juice of the mature canes of frost-free regions of the south was used to make "white" and "rock" sugar. The juice was carefully boiled until the strike point then set aside to allow crystals to form. In the next stage, the crystals, now drained of molasses, were packed into inverted earthenware cones and "clayed" in a manner reminiscent of colonial America to produce the "white" sugar. Sung distinguishes five grades of "white" sugar. The whitest sugar was, of course, at the top of the cone and Sung (1966: 128) refers to it as "western sugar" because sugar imported to China from western countries – India and Persia, presumably – was white and of high quality. Sung's translators (1966: 132) in a footnote that is not entirely clear in its meaning, appear to equate this "white" sugar with the "sugar frost" of the early Chinese industry, and indeed if they are correct in this association, "claying" in China long antedates its use in the colonial American sugar industry. Where and when "claying" was first used to whiten sugar and whether or not the knowledge diffused around the sugar-making world or was independently discovered in several places are questions for which at the moment there are no certain answers. "Rock" sugar was made by dissolving "western" sugar, adding albumen, boiling and skimming off impurities until at the correct moment split pieces of new green bamboo were thrown into the syrup. As the syrup cooled crystals, or "rocks," formed around the bamboo. Finally, Sung described the making of the popular animal-shaped candies. Clarified syrup was poured into cold molds; a layer of sugar coated the interior of the molds and the surplus syrup drained out. The delicacies made in this way were hollow (Sung 1966: 124–32).

Descriptions of sugar production in China c. 1900 show that very little if any change in milling technology or methods of manufacture had taken place in the three centuries since Sung wrote. Simple two-roller mills, with either wooden or stone rollers, were still the standard means of extracting the cane juice and some very good photographs survive (Hommel 1969: 111–14; Pitcher 1909: 49). However, Baxa and Bruhns (1967: 218) reproduce an illustration of milling in late nineteenth-century China that shows what are clearly three-roller mills, a hint perhaps of the diffusion of technology back from the Americas. The procedures and terminology in Hommel's accounts of sugar making are strongly reminiscent of Sung's. One recent foreign intrusion was the construction of modern refineries in Hong Kong and Swatow to satisfy the tastes of the increasingly numerous western population in China. They were built by the well-known Hong Kong merchant houses of Jardine Matheson, Butterfield and Swire (Allen and Donnithorne 1962: 174).

The Chinese sugar industry for several centuries before 1914 can be characterized in the following way: it was a labor-intensive peasant activity with a simple technology; the peasants did not adopt new "better" technology from the western world; it served for the most part domestic markets; the relationships between laborers, landowners, merchants and consumers remained stable. Explanations for this continuity can be given in such terms as land tenure, the availability of capital, the low purchasing power of the consumers. But the matter can be seen as part of the broader inquiries into why modern China failed to generate either a scientific revolution or a capitalist system. Until 1600 or so, China was technologically in advance of the West – this general statement holds true in the specific case of sugar mill technology – and thereafter did not follow the new course of economic development charted in the West. Bray (1984; 1986) has argued that the reasons for this lie in the conditions of wet rice cultivation, which requires large amounts of labor rather than capital investment and machinery, and leads to "petty commodity production" rather than capitalist enterprises. "It is our contention, then," writes Bray (1984: 615) "that while rural societies economically dependent on wet rice cultivation are not incapable of mechanization and industrialization, they are unlikely to engender these processes simultaneously." Bray sees two models of modern historical change, a western one and that of the wet rice economies. Sugar cane in China was a minor crop in a society dependent on wet rice cultivation, and sugar production shared the characteristics of the society of which it was a part.

Southeast Asia

Sugar cane was a crop of long-standing domestic importance throughout Southeast Asia before the first European merchants appeared in the region, and a few places such as southwestern Taiwan produced a surplus for export. During the years of colonial administration, the production of sugar in Java, Taiwan, the Philippines and Malaya was transformed, commercialized, and the four countries became significant exporters. There are a number of elements in the explanation for this striking contrast with the sugar industry in India and mainland China: one is ease of access to markets, given that Southeast Asia lies across one of the world's mainstreams of commerce and that the region's islands and peninsulas are readily accessible to shipping; secondly, the small size of the individual territories meant that the foreign authorities could more effectively enforce agricultural as well as other policies; and, as a further consideration, in the second half of the nineteenth century a new market arose for sugar in Japan which Southeast Asia was ideally placed to supply.

The demand for sugar in Japan was to lead during the late nineteenth and early twentieth centuries to the transformation of sugar production in Taiwan and to make Taiwan into a virtual sugar colony. In Japan, as in the West, industrialization led to an increase in the per capita consumption of sugar – from 5 to 12 pounds during the years 1888–1903 (Williams 1980: 225–9) – and what had once been a luxury handled in small quantities became an item of common use. Before the Meiji Restoration (1867), Japan had imported sugar variously from China, Taiwan and the Dutch in Java as well as supporting its own small industry in the far southwest of the country, on the islands of Shikoku and Kyushu. The town of Kagoshima, on Kyushu, was the refining center, and it drew not only on the neighboring districts for raw sugar but also on the Ryukyu Islands. In the years before the Restoration, this industry was enjoying a period of expansion and prosperity with some sixty ships employed in carrying the sugar to Osaka, the main distributing center. One consequence of the opening of the ports to foreign trade after 1867 was the destruction of this industry as it could not compete with the cheap imports (Hirschmeier 1964: 82, 102–3). The growing drain of these imports on foreign exchange became a concern to the government and no doubt the potential of nearby Taiwan as a producer of sugar was a factor in the decision of the Japanese to occupy it in 1895.

Before the Japanese occupation, Taiwan had a traditional sugar industry, very similar to that on mainland China. When sugar manufacture first began on Taiwan is unrecorded, but a small amount of sugar was being made when the Dutch arrived in 1624. The Dutch during their brief rule of a small enclave on the southwest coast promoted sugar cane cultivation and succeeded in raising exports from perhaps 80–200 tons in 1636 to 1,300 in 1660. The Japanese appear to have been the most important customers. In 1662, the Cheng family, refugees from the new Manchu dynasty on the mainland, drove the Dutch out, but they continued the Dutch policy of encouraging the sugar industry with the result that exports had reached an annual 18,000 tons by the time the Manchus gained control of the island in 1683. The next two centuries of the Taiwanese industry are poorly documented but north China was a major market and also Japan. Exports do not appear to have increased until after 1870 when for a decade there were new markets in Australia and California until protective tariffs closed them. By the 1880s, Japan was again the most important market for Taiwanese sugar (Williams 1980: 220–3).

There were then possibly 1,400 simple inefficient mills on the island, and the cane was grown by small-holders and tenants in a variety of contractual arrangements with moneylender-millers and landlords. There were also cooperatives in which farmers worked together to harvest and mill the cane. The Japanese government gave the task of modernizing this industry to Japanese entrepreneurs and encouraged them with tax benefits, assistance in acquiring land, in building sugar mills and narrow-gauge railways and, very importantly, guaranteed them a protected market for their sugar in Japan. The Formosan Sugar Company, organized in 1900, was the first of the sugar companies and included members of the imperial household among its shareholders. In the southwest of the island, in the area of the former Dutch enclave, the Japanese established a sugar cane research station which was to become one of the foremost in the world. After 1909, the various companies worked through the government-sponsored Sugar Industry Association of Taiwan to reach agreements on the prices to be paid to the farmers for cane, on the supply of cane to the mills and on the marketing of sugar. Thanks to this collusion and to their economic and administrative power, the companies found that cane-farming was a cheaper means of obtaining cane than through the alternative of establishing plantations. In theory, a farmer had the option of refusing to grow cane but companies could secure compliance from farmers by a variety of means: loans to farmers, the leasing of land and the supply of irrigation water could all be made conditional on the cultivation of cane. The Japanese rapidly reduced Taiwanese participation in the industry to small-scale cane-farming. The traditional mills disappeared, and by 1915, thirteen Japanese companies owning thirty-five modern central factories dominated the industry (Williams 1980: 229–35, 244).

Java was the scene of a rather different type of colonialism, but one that nevertheless converted the island during the course of the nineteenth century from a minor producer of sugar into a major sugar colony, a transformation that represented for the colonial power, the Dutch, an enormous success. Whereas in the early 1820s exports of sugar were a mere 1,000 tons or so annually, by 1920 they had reached more than 1.5 million tons and the revenues they generated exceeded in value the revenues of all other exports of Java combined (Furnivall 1939: 337). The Dutch first exploited Java through their East India Company (1602–1800), and then administered it directly through government-appointed officials. The Dutch were mercantilists in their approach to empire, looking on their eastern territories as suppliers of agricultural goods to be sold on the world markets for the benefit of Holland. As Holland did not become an industrial country comparable to Britain during the eighteenth and nineteenth centuries, the encouragement of a consumer market for industrial goods in its eastern territories would not have served its interests because it could not supply the goods but rather those of Britain and, later, of Japan who could. Instead, the Dutch sought to increase the cultivation of export crops at the expense, if necessary, of the peasant standard of living. Through various controls, low wages and low rents they forced the peasants to subsidize the export economy. The poor returns given the cultivators were also a means by which the Dutch tried to compensate for their inability to provide adequate capital for the development of their eastern empire.

From the mid-years of the seventeenth century, the Dutch East India Company encouraged the cultivation of sugar cane in the western part of Java, immediately around their capital of Batavia (Djakarta), and monopolized the export trade until 1809. Given there was an internal demand for sugar, particularly to make arak, relatively little was available for export, perhaps 6,000 tons annually during the better years of the eighteenth century. Some Javanese sugar did reach Europe, but the length of the ocean voyage put it at a disadvantage in the competition with West Indian sugar; most of it was sold to markets within southern Asia. Despite the Dutch overlordship, the Javanese industry owed more to China than to Dutch experience of sugar production in Brazil and the West Indies. The Dutch did provide much of the finances and there were some Dutch planters, but essentially the industry was in the hands of Chinese entrepreneurs, and the technology was Chinese. The two-roller mill, stone or wooded, and turned by oxen, was the standard mill, and, as in China, West Indian technology appears to have had no impact. As late as 1829, there were said to be only five "West Indian mills," i.e., three-roller mills, on Java. The Chinese occupied the skilled positions, while Javanese field hands were recruited from Cheribon and points further way along the north coast. These immigrants apparently made more tractable workers than the Javanese who lived within the immediate vicinity of Batavia. By the early years of the nineteenth century, this industry was in difficulties, not only undercapitalized and technologically backward, but over the decades it had eroded its resource base, depleting the forests and exhausting the soils, so that any renewal of the Javan industry would have to take place elsewhere (Furnivall 1939: 40–1; Knight 1980: 177–81).

Governor General Johannes van den Bosch who was appointed in 1830 is credited with the development of the export economy of Java, and his policy that brought this about is generally known as the Culture System. In basic outline, the System was very straightforward, although in actual operation, depending on crop and region, it could become quite complicated. The government ceased to collect the taxes formerly paid by the peasants on their land but in return the peasants had now to devote part of their land to the cultivation of an export crop designated by the government or work a given number of days a year on government projects. The government, moreover, co-opted the traditional social structure in support of the System, obliging the indigenous elite to organize in an appropriate manner the allocation of village land and labor to the cash crops and to oversee the work of the peasantry. At various stages in the growing and processing of a crop, the government could and did interfere to manage quantities and prices, and it established for itself a virtual monopoly of the export trade. Cotton, tea, indigo, tobacco, pepper, cinchona, cinnamon, cochineal and silk were all encouraged at one time or another under the System, but the greatest success came with coffee and sugar (Elson 1984: 32; Geertz 1968: 52–5).

The new sugar industry evolved along the north coast of Java and in the east of the island, areas with a pronounced dry season, but with fertile soils and a labor force. Here both rice, the staple food, and sugar required irrigation, and under the Culture System, the peasants had to devote part of their paddies to the cash crop. More rice now had to be extracted from less land, and labor devoted to cultivating sugar, to carting and milling it: an immediate consequence therefore of the Culture System for the Javanese rice-farmer was an increase in his work load. About one fifth of a village's paddy was planted to cane. Each factory was assigned a circular supply area, with a radius of about 8 km. The government advanced money to Chinese and Europeans to build modern mills and factories, and supplied them with cane, the fuel and the workers. The factory owners, in return, were under contract to deliver the sugar to the government docks. Throughout the period of the Culture System (1830–70), there was a basic continuity in the organization of sugar production, and the effectiveness of the System in generating sugar exports can clearly be seen in Table 8.1 (Elson 1984: 36, 48–50, 72–6, 122–3).

Geertz (1968: 63, 65) has written that the Culture System "generally substituted the labor of the Javanese for the capital Holland lacked in

The sugar cane industry

Table 8.1 *Sugar exports of Java*

	Tons		Tons
c. 1750	6,000	1860	128,536
1808	5,876	1875	209,000
1815	1,235	1880	222,000
1825	988	1890	367,000
1830	6,670	1900	736,000
1840	63,187	1915	1,348,000
1850	85,414	1920	1,588,000

Source: Furnivall 1939: 75, 104, 129, 171, 207, 316.

laying the preparative foundations of a rapidly accelerating, if distorted, process of economic growth" and it "represented an attempt to raise an estate economy by a peasantry's bootstraps." The very success of the Culture System now led not only to demands, but indeed to a need, for change in the regulation of the industry. In the decades since 1830, the System had built up the infrastructure of an export industry with irrigation networks, roads, modernized factories and so had created an environment in which private entrepreneurship was much more likely to flourish. The increasing size of sugar factories also called for new patterns of organization. With two laws passed in 1870, the Sugar Law and the Agrarian Land Law, the government began gradually to "privatize" the industry, a process which took until the early years of this century to complete. The Sugar Law ordered a gradual reduction in the area of cane cultivation under government management (Elson 1984: 127) and its efficacy is reflected in the fact that privately cultivated cane increased from 9% of the total in 1870 to 97% in 1890 (Geertz 1968: 84). The Agrarian Land Law (1) accepted a long-standing argument that uncultivated "waste land" belonged to the state and could be leased for private plantations, and (2) permitted the Javanese peasantry to rent, but not to sell, land to foreigners for the establishment of plantations. In effect, the Agrarian Land Law was a tactical means of protecting the basis of village society while opening up land to commercial interests. With these dispensations, entrepreneurs could plan and manage all aspects of production from cultivation of the cane through milling and manufacture to selling the product on the international market. The mercantilism that had marked the Dutch approach to sugar production in Java since the seventeenth century gave way to corporate capitalism (Elson 1984: 127–8; Geertz 1968: 65–6, 83–4).

This switch in form of ownership was not peculiar to Java but part of a general trend in many regions of commercial cane sugar production during the late nineteenth and early twentieth centuries: as already noted the industry in the Caribbean was passing at this time into the control of corporations in London or New York. The takeover of the industry in Java by corporate capitalism was aided during the 1880s by the low price of sugar. The entrepreneurs who had emerged during the period of the Culture System and who were the first beneficiaries of the policies introduced in 1870 did not have the resources to tide them over the crisis and sold to large corporations that were for the most part based in Holland. Concentration of ownership continued in subsequent years, and by 1915 one company, the Nederlandsche Handel Maatschappij, originally a shipping company founded in 1824, controlled as many as thirty-eight factories (Geertz 1968: 84–5). In a second way – by relying on the breeding of new varieties of cane – the Javanese industry paralleled developments elsewhere. In 1887, botanists in Java managed to grow cane from seed and this led to a very successful breeding program at the East Java Sugar Research Institute, producing the POJ (Proefstation Oost-Java) canes. Between 1840 and 1910, the average yield of sugar per hectare in Java rose from 2 tons to 10. While part of this increase in yield can be attributed to improvements in irrigation and the abandonment of less fertile soils, it was due in large measure to the introduction of the better POJ varieties of cane (Cramer 1952: 143–4). The Javan industry remained distinctive, however, in one important aspect, and that was its relationship with paddy rice cultivation. To quote Geertz (1968: 86), "A village, sometimes willingly, sometimes coerced by its leaders and local civil servants, contracted a 21 $\frac{1}{2}$-year lease with an estate," and sugar cultivation existed in a complex rotation with rice on the village lands (Geertz 1968: 88; Elson 1984: 180).

British rule in Malaya provides a very different example of colonial agricultural policy than that followed by the Dutch in Java. The British only gradually gained possession of the Malay peninsula during the course of the nineteenth century. In 1786, they took the island of Penang and in 1800, began to administer a strip of territory, the new Province Wellesley, on the mainland across from Penang; in 1819, they added Singapore Island and Malacca five years later. There were no further territorial acquisitions until 1874 when they reversed their policy of non-intervention towards the various Malayan states on the peninsula by extending their "protection" to Perak, Selangor and Sungei Ujong. Yet more states were gathered in to form the Federated Malay States in 1896. The British, like the Dutch, wanted to make of this new possession a source of tropical crops for export, but rather than force the indigenous population to cultivate the cash crops and force also the indigenous elite to administer the work as the Dutch were doing in Java, the British adopted the very different approach of importing both planters and workers. The planters from Europe would bring capital and enterprise and, so the reasoning ran, British India would provide the cheap labor that the sparsely populated Malay territories could not themselves supply. The colonial government hence introduced the appropriate immigration and labor regulations, provided land for plantations at nominal cost and built roads and railways. The European planters tried various crops with little success, despite the support the state was giving them, and while they blamed the external world for their difficulties, a basic cause lay in themselves for many were new to Malaya, even to agriculture, and had neither the capital nor the knowledge to convert tropical forest into successful plantations. Another group of immigrants, the Chinese, were in fact rather better at the job, prospering in the cultivation of spices and manioc for the manufacture of tapioca. During the second half of the nineteenth century a modern sugar industry emerged but it did not retain government support and gave way just before 1914 to rice paddies, coconuts and rubber (Ghee 1977: 14–16, 43; Jackson 1968: 133–8).

Malaya, before the arrival of the British, was part of the Chinese sphere of the sugar industry along with pre-Japanese Taiwan, pre-Dutch Java and south China itself. The small fields of cane, the two-roller mills, the simple means of manufacture and poor quality sugar elicited the same disapproving comments from British officials as they had elsewhere in the Far East from other Europeans with some knowledge of the West Indies. Nevertheless, despite the shortcomings of their technology, the Chinese began to develop the industry in a very modest way in the security of British-administered Province Wellesley. In the late 1830s, European planters arrived on the scene and their presence no doubt helped the lobbying that in 1846 resulted in Malayan sugar being allowed to enter the home market on the same basis as West Indian. However, Singapore Island, although part of the empire, was excluded from this privilege for fear that it would become a conduit by which sugar from non-British parts of Southeast Asia reached Britain. This exclusion effectively put an end to commercial sugar cultivation in the vicinity of Singapore. Until 1874, the Malayan sugar industry was concentrated in Province Wellesley (Table 8.2). Gradually, the Chinese and European planters came to rely on each others strengths. The Europeans had the capital to build modern mills and factories, but were not as successful at cultivating cane with their Indian laborers as the Chinese planters who relied in the main on Chinese laborers. Over the years, the Chinese abandoned their two-roller mills and entered into contracts to supply the European mills with cane. Expansion was interrupted in the crisis of low prices of the 1880s. Many laborers, Indian as well as Chinese, left the cane fields for other parts of Malaya as well as for Sumatra, but there was a recovery, helped by the British who further facilitated the

Table 8.2 *Sugar exports of Malaya*

	Tons		Tons
From Penang (Province Wellesley)			
1840	650	1873	10,600
1845	750	1883	23,000
1850	2,700	1889	19,500
1855	4,000		
From Perak			
1883	5,750	1902	25,500
1890	5,500	1910	9,500
1900	17,000	1913	300
From Perak and Province Wellesley			
1900	36,000		

Source: Jackson 1968: 167 and figs. 26, 27 and 29.

immigration of laborers from India, and the industry lasted for another twenty years (Jackson 1968: 128–54).

The planters of Province Wellesley had long known that, immediately to the south, the alluvial coastlands of Perak were very suitable for sugar cane cultivation but they did not begin any developments there until after the British had brought some administrative stability to the state in 1874. The Chinese were the first to establish plantations there and, having learnt from the British, set up modern mills. British enterprise was largely in the hands of two corporations, the Perak Sugar Cultivation Company and the Straits Sugar Estates Company. By 1900, Perak and Province Wellesley together were producing some 36,000 tons of sugar a year which ranked it as a more important sugar colony than many Caribbean islands. The industry was characterized by large plantations with foreign labor, management and capital (Jackson 1968: 154–67).

Yet the British for several reasons allowed this sugar industry to disappear. A prime consideration was the fact they already had an abundance of sugar in subsidized beet from central Europe and from their West Indian colonies. They had no need for Malayan sugar. Moreover, given the overproduction of sugar and the great competition in the international market, there was little likelihood of sugar production in Malaya becoming very profitable in the short term. There were also local problems. The industry had always experienced difficulties in keeping an adequate labor force, in large measure because it could not afford to pay good wages, and these problems simply were exacerbated as the Malayan economy developed and in the process drew laborers away from the sugar plantations to different types of work. There was competition for land from other export crops. In particular, coconuts and the new plantation crop, rubber, held promise of far better profits, and landowners began to turn their resources to cultivate them. A further important cause of the decline was the British wish to encourage rice cultivation in order to feed the growing population of Malaya. An ideal location for paddy fields was coastal Perak into which also sugar cane cultivation was expanding and by the 1890s, the British had come to view these two forms of land use as incompatible. In newly irrigated areas, they adopted a policy of protecting the paddies from sugar planters, a policy that severely limited the planters' access to land. Without government support, the sugar industry could not overcome the other difficulties it faced and by 1914, sugar cane in Malaya had ceased to be a crop of economic importance (Ghee 1977: 30–1, 42–5; Jackson 1968: 167–75).

The Philippines in the 1830s exported annually about 15,000 tons of sugar, an amount that had increased to 236,000 by 1914. However, in this instance the colonial power, Spain, can claim very little responsibility for the development of the industry as it was too poor to provide capital for improvements and its own small market was already glutted with sugar from Cuba. The growth of the industry was a result of the increasing consumption in the world at large and the pattern of the Filipino trade shifted in responses to changes in supply and demand in the world's sugar economy. Britain remained a major purchaser throughout the nineteenth century, taking 110,000 tons in 1881 and nearly that amount on three occasions in the 1890s, but between the Spanish-American War and 1914 Britain lost interest in Filipino sugar, in some of these years buying none at all, a consequence perhaps of its renewed concern for the welfare of its Caribbean colonies. The United States, too, was a major purchaser throughout the nineteenth century, and the importance of this market for the Philippines greatly increased when the United States became the occupying power. In 1909, the United States agreed to admit 300,000 tons of Filipino sugar duty free, and in 1913 even this quota was eliminated. But the Philippines also found markets for the sugar closer to hand. Australia was a major outlet until the 1880s when its own industry came on stream. Japan imported some over the years, but the most important Asian market after the 1880s was China which during the first decade of the twentieth century took more than half the total of Filipino exports (Larkin 1984).

This export industry depended in 1888 on 5,000 animal-powered mills, 35 water mills and only 239 steam-powered mills. Milling in Luzon, where oxen and water buffaloes turned old stone or wooden rollers, was reminiscent of China. It produced some clayed sugar, known locally as *pilon*, but low quality muscovado sugar made up by far the greater part of the exports. By the end of the nineteenth century, the Filipino industry was backward, more so indeed in the opinion of Baxa and Brunhs (1967: 223–4) than that of any other European colony, a statement which should be qualified to exclude India. The increase in exports was made possible by a great extension in the area cultivated in sugar cane, and central Luzon, Negros and Cebu were drawn into the industry. The landowners were Filipinos although in some areas Spanish families were prominent; the labor force also was found locally. Foreigners, however, managed the export trade, and foreign, that is American, capital financed the modernization of the industry that began to get underway after the Spanish-American War.

Not surprisingly, given the markets for Filipino sugar, the principal merchants' houses involved in the export trade were British, American and, increasingly towards the end of the nineteenth century, Filipino Chinese as they had contacts in and knowledge of their home markets. They also helped the more enterprising local planters to finance steam-powered mills and even a few vacuum pans. With the United States governing the Philippines, and once Filipino sugar had been given easy access to the American market, American capital flowed into the industry. By 1914, three large American-financed central factories were in operation, and more quickly followed so that by 1925 there were thirty-four modern factories. Some of these factories had been built by Filipino investors, but, nevertheless, in the Philippines as elsewhere during the modernization of the sugar industry ownership passed to foreigners (Baxa and Bruhns 1967: 224; Larkin 1984).

Taiwan, Java, Malaya and the Philippines thus provide an interesting study in imperial policy towards sugar production in Southeast Asia. In all four cases government support, even control, was crucial to the transformation. Empire, market and finance were linked. In the Philippines, modernization was delayed until a weak and declining imperial power was supplanted by a dynamic one that could provide both capital and a market. The transformation in Taiwan followed quickly the Japanese occupation and, as Japan's only sugar colony, essential to the success of its policy of reducing dependence on foreign suppliers, the government ensured that the capital flowed in to exploit the island's potential. Java likewise was Holland's most important source of sugar and given the extent of its Indonesian possessions it could develop there the production of both sugar and rubber. In consequence of this imperial activity in Southeast Asia, old small-scale industries that owed a good deal to Chinese influence in technology and organization were replaced by western-style central factories owned by large corporations.

9
The Indian Ocean and Pacific colonies: 1800–1914

In the nineteenth-century scramble for empire in Africa and the Pacific, the acquisition of more sugar colonies was not among the motives of the imperial powers: sugar cane was, after all, already being grown in abundance. Yet, despite the low prices and competition from beet, in some of the new colonies sugar cane proved to be more profitable than other crops, and so a new group of sugar colonies came into being. Each of these colonies evolved its own distinctive personality in which geographical circumstance, date of entry into the sugar business and policies of the colonizers were formative factors. On Mauritius and Réunion, the industry was established by the French early enough to have been based on slave-worked, family-owned plantations and so, as in the West Indies, it had to adapt during the nineteenth century to emancipation and the central factory system. In Natal, Fiji and Hawaii, commercial sugar production had begun so recently that large factories were part of the initial structure but the planters did have to reach a *modus vivendi* with the indigenous people. Nevertheless, there is inevitably much that is common to the human geography of each of these colonies, given that they were all in the same line of business and had to adopt the same innovations in technology, plant breeding and financial organization in order to remain competitive. In Queensland, a very distinctive sugar industry began to emerge at the time of the formation of the Commonwealth of Australia in the years of this century, one that relied on European labor to work the cane fields. It deserves separate consideration.

The new colonies

A market for their sugar was the prime requisite for all these new colonies – without one they would neither have entered nor continued in the industry – and the markets in turn influenced the course of development of the colonies they supported. This latter point is clearly shown by the experience of the two Indian Ocean islands which became divided in their imperial loyalties once Britain had taken Mauritius as part of the spoils following the Napoleonic Wars. They shared some difficulties: the serious outbreak of disease in the cane in the 1860s was followed by a decline in production (Fig. 9.1), and the opening of the Suez Canal in 1869 meant that rather than being on the major shipping route between India and Europe they were now quite literally in a backwater, but Mauritius recovered rather better than Réunion. Mauritius did have the advantage of a reliable supply of cheap labor from British India but very importantly it did not have to compete with a protected domestic British beet industry. Britain, although an importer of beet sugar, did not grow it in any significant quantities while France did. Indeed, there was a surge in beet production in France after 1865. To make matters worse for Réunion, during the last quarter of the century French imperial interests in the Indian Ocean turned away from what was now a colony devoid of any strategic significance and which could not promise to bring more wealth or glory to the recently acquired, altogether more intriguing, possession, the huge island of Madagascar (Toussaint 1972: 245–9; Scherer 1980: 70–3, 80, 88). By 1914, Mauritius was exporting four times more sugar than Réunion.

Unlike the two islands, Natal had the benefit of an expanding domestic market. Natal's main competitor was Mauritius, not only in Natal itself, but also in Cape Province and in the mining populations of Kimberley and the Witwatersrand. At first the sugar planters in Natal could only arrange tariff protection in their own colony but as Natal entered customs unions with its neighbors and then became part of the Union of South Africa, the scope of this protection was greatly increased (Richardson: 1984: 241, 248–9) and the future of its industry guaranteed.

The Pacific colonies were far too remote from Europe to hope to compete in the sugar trade there and turned instead to new markets. Fiji at first exported sugar to Australia and New Zealand, but as the emergence of Australia's own industry gradually reduced the importance of that market, exports to Canada increased (Moynagh 1981: 5). Hawaii was part of the American empire and sold its sugar in the United States. It is perhaps not too great a simplification to state that the sugar industry either made or ruined Hawaii depending on one's point of view. A white elite descended from American Protestant missionaries knew that the sugar industry it had founded in the 1850s and 1860s could not flourish without being able to export to the United States, and in working to achieve this, the elite undermined the precarious independence of Hawaii. It found allies in the United States among those who recognized the importance of Pearl Harbor in the geopolitics of the Pacific, opponents in Louisiana planters who feared more competition, and in the final dénouement strategic imperatives overcame Southern political clout. By the Reciprocity Treaty of 1876, the United

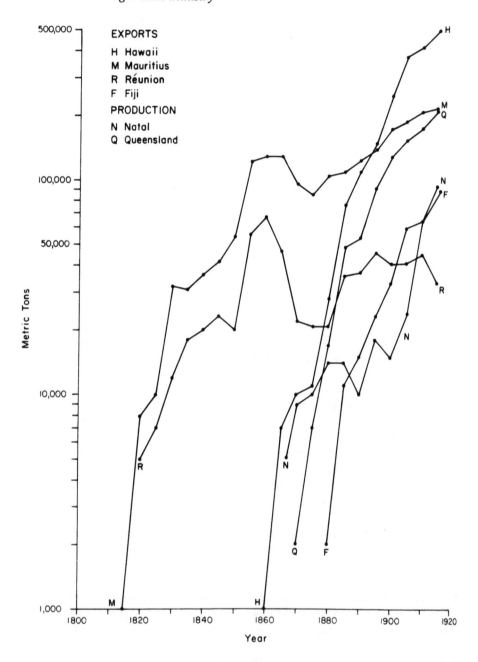

States allowed Hawaiian sugar to enter duty free in return for a guarantee that Hawaii would not make land grants to any other country. The Treaty was twice renewed, always amid controversy, and the debate over the future of islands dragged on both in the United States and Hawaii until 1898 when U.S. naval action in the Philippines at the outbreak of the Spanish-American war emphasized in a very dramatic way the pivotal role of Pearl Harbor in the control of the Pacific. Within weeks Congress voted for annexation (Bell 1984: 8, 22–4, 30–7). The export figures from 1876 on (Fig. 9.1) clearly demonstrate the fundamental importance of the American connection to the Hawaiian sugar industry. By 1899, sugar accounted for 96% of the total value of Hawaii's exports (Tate 1968: 121).

The most visible similarity between these colonies, apart from the fields of cane, was the multicultural nature of their societies. The plantations were, of course, as elsewhere in the colonial world, an exotic intrusive form of land use that brought with them an expatriate population of owners, managers and laborers. The planters and managers in these colonies were usually French, British, American or Australian. The fact that the colonies either could not or would not provide labor for the plantations led to an international migration on a far greater scale than that prompted by the need for managerial talent.

There was indeed no alternative to immigrant labor on Mauritius and Réunion which were uninhabited at the time of their discovery by Europeans. The first slaves were Malagasies but later the French took slaves at several places along the east coast of Africa. Possibly as many as 100,000 were brought to Mauritius over the years and at emancipation in 1835 some 70,000 slaves received their freedom; the 60,000 slaves of Réunion had to wait until 1848 for freedom (Toussaint 1972: 76, 206, 209). As in the West Indies, abolition brought complaints from planters about the unwillingness of the ex-slaves to work and they looked to the imperial governments to provide new sources of workers; others saw immigration as a means of flooding the labor market and thereby lowering the cost of labor. Certainly, the ex-slaves received little incentive to remain on the plantations and many drifted off to become fishermen, artisans of laborers in the towns. Mauritius immediately turned to India and for nearly a century indentured Indians came in very large numbers (Table 9.1) at first to work and then to settle permanently. By 1872, they had completely replaced the ex-slaves in the cane fields. The French on Réunion did not

Fig. 9.1 Sugar production in the Indian Ocean and the Pacific colonies, 1800–1914 (five-year averages)

Sources: Fiji: Moynagh 1981: 47. Hawaii: Colson 1905: 72; Deerr 1949–50: vol. 1, 258. Mauritius: Deerr 1949–50: vol. 1, 203–4. Natal: Richardson 1984: 251. Queensland: Graves 1984: 264. Réunion: Colson 1905: 368; Scherer 1980: 54, 80.

Table 9.1 *Indentured Indian laborers*

Fiji	1879–1916	60,553
Natal	1860–1911	152,814
Mauritius	1836–1909	450,000
Réunion	1861–85	117,813

Note: many returned to India on the completion of their contracts.
Sources: Fiji: Moynagh 1981: 21; Natal: Richardson 1984: 241; Mauritius: Toussaint 1972: 212; Réunion: Scherer 1980: 74.

have such easy access to British Indian workers. After negotiations, the British finally agreed in 1861 to permit the French to recruit Indians but imposed many protective clauses about the treatment of the laborers. When in 1884, the British consul in Réunion considered these conditions were not being met, the British canceled the agreement. During the last years of the century, the planters of Réunion were recruiting Mozambiquans, Somalis, Comoro Islanders, Yemenis and Tonkinese (Scherer 1980: 64–5, 76; Simmons 1982: 24).

In Natal, Fiji and Hawaii, the indigenous population had its own land, agricultural economy and even some political power which meant that it was in a position to refuse to work on the sugar estates if it so wished. In Natal, the attempts to woo the Africans to work on plantations failed (Richardson 1984: 241). The Natal government following well-established precedent turned to another part of the British empire, India, to recruit indentured laborers. In Fiji, the tragic decline in population that followed the introduction of alien diseases and of alcohol meant that the survivors in the late nineteenth century had abundant land on which to sustain themselves; moreover, calculations by colonial officials showed that by working this land reasonably industriously, Fijians could support themselves in a far better way than by working for the low wages offered by the plantations. For the colonial government to have exacted labor for the plantations against the wishes of the Fijian chiefs would have required a level of compulsion it was neither willing nor able to impose. Rather, the government began a policy of protecting Fijian agriculture and Fijian land rights. There was a brief unsatisfactory experiment with laborers brought from other Pacific islands then, in 1875, at the suggestion of the governor, Sir Arthur Gordon, who had experience of both Trinidad and Mauritius, the planters began to recruit indentured Indians in a fateful decision that was to solve the immediate labor problems of the sugar industry but at the same time

Table 9.2 *Immigrants arriving in Hawaii, 1852–99*

Ethnic group	No.	Period	Percentage of total
Chinese	56,700	1852–97	39.0
Japanese	68,279	1885–99	46.7
Portuguese	17,500	1878–86	12.0
Pacific Islanders	2,500	1878–85	1.7
German	1,300	1882–5	0.8
Total	146,279		

Source: Beechert 1985: 86.

to lay the basis of political ones that survive to this day (Moynagh 1981: 15–21).

Matters evolved somewhat differently in Hawaii. Planters complained of the high wages the native Hawaiians demanded and discussed ways in which they might be compelled to work in the cane fields, but at mid-nineteenth century to enslave or in some other way to coerce work from the citizens of an independent country in which Christian missionaries were a very strong presence was clearly out of the question. But despite the rhetoric of the planters, the records show that Hawaiians did work if offered a good wage, and indeed until the great expansion of the sugar industry that followed the Reciprocity Treaty of 1876, the majority of the sugar workers were native Hawaiians. After 1876, other sources of labor had to be found because the number of Hawaiians was now so reduced by the effects of alcohol and introduced disease – down by 1860 to 22% of the 1778 population – that even if all of them had abandoned other activities to work in the sugar industry the numbers would not have been sufficient. Inevitably, planters' thoughts turned towards India, but Indians were not to come to Hawaii: the British government was now very sensitive to the conditions in which the indentured laborers worked and was very cautious about extending the system to territories it did not control. The answer lay in China and Japan, and in the years before annexation these two countries provided the great majority of the immigrants to Hawaii. This success did not leave the American elite entirely comfortable; deeply racist in its thinking it attempted to Europeanize the labor force by importing Portuguese at considerable expense (Table 9.2). Despite these large numbers, the labor supply continued to preoccupy the industry because so many of the immigrants either returned home or moved on to California. Between 1905 and 1916, for instance, 104,406 immigrants arrived but 83,451 left for a net gain of only 20,955. After annexation, Japan and Portugal continued to provide the bulk of the immigrants, but now also Filipinos, Koreans, Spanish and Russians arrived to add to the ethnic diversity of Hawaii (Beechert 1985: 60, 64, 85–8, 132).

In all these colonies, as elsewhere in the cane-growing world, the introduction of the central factory system during the half century leading up to 1914 was accompanied by a concentration of cane land into huge estates and by the transfer of the ownership of the industry to large corporations. The process of adoption of the new system in Mauritius and Réunion had more in common with the West Indies than with Fiji, Natal and Hawaii.

On both islands there was a strongly entrenched planter class which had no wish to give up its autonomy or suffer a loss in prestige and social rank that the closing down of mills would entail. Nevertheless, few of the owners of the mills could expect to become owners of central factories. In the struggle to modernize, some planters inevitably overextended themselves, fell heavily into debt and had their property acquired by banks or by more successful planters; others managed to stay in the business of milling cane but through building small factories that would soon become uneconomic. By the early 1900s, there were still sixty factories on Mauritius where there had been three hundred mills and thirty-four on Réunion. They varied greatly in size: on Réunion the smallest had a capacity of 300 tons of raw sugar a year, the largest 3,000 (Colson 1905: 369; Mannick 1979: 78–9). In Réunion, the process of concentration of landownership took on a particularly dramatic aspect. One family, the Kervéguen, amassed 30,000 hectares, and in order to pay their many employees more easily received in 1854 permission from the island Council to issue their own currency, which turned out to be 800,000 recently demonetized Austrian coins. These became know locally as "Kervéguens" and circulated for twenty years, guaranteed by the fortune and immense prestige of the family until the island became part of the French monetary system. A second very large holding on Réunion was the work of a French bank, the Crédit Foncier, which organized the 9,000 hectares of thirty-seven bankrupt plantations along with twelve factories into a modern company, the Sucreries Coloniales (Scherer 1980: 77–9). There were also countervailing trends that led to considerable diversity in the pattern of land tenure on both islands. Some planters in need of capital sold all or sections of their land in parcels of varying sizes to an emerging class of cane-farmers; others introduced *métayage*. *Métayage* contracts were usually verbal, informal, but characteristically required the tenant to grow cane which would be divided in agreed proportions with the landowner. Such a contract was a means by which a planter in times of scarce labor could fix workers to his land so as to guarantee a supply of cane for his mill; the tenant in turn, owning a portion of the crop, benefited directly from his own labor. *Métayage* increased greatly in Réunion after Britain stopped the flow of Indian indentured laborers (Toussaint 1972: 262–4). The rationalization of the milling and manufacturing on the islands was by no means complete in 1914, but already the introduction of the central factory system had complicated the patterns of land tenure and social structure.

In Natal, the transition from a domestically owned, planter-based industry to a central factory system owned by large corporations took about thirty years to complete, beginning in the 1870s. As well as the economic pressures to adopt the new technology, there was in Natal an additional incentive to modernize: the failure of the Green Natal variety of cane in the 1880s and its replacement by the Uba variety which was then grown successfully for fifty years. Uba had a hard, tough, rind resistant to disease and rodents but required milling in a powerful mill. Also, its relatively long growth cycle meant a planter who wished to maintain the same flow of cane to the mill as before had to increase his acreage of cane. Thus Uba favored large estates, and required modern mills (Barnes 1974: 23; Richardson 1984: 253). The consolidation of land-holdings resulted in estates of many thousands of acres, and the factories in the usual pattern became fewer but larger. The average mill in 1874 had a capacity of approximately 130 tons of raw sugar a year; in 1898, 700 tons and in 1909, 1,625 tons. An increasingly prominent figure on the scene was the "miller-cum-planter," in the parlance of the Natal sugar industry, who owned a factory, grew his own cane but also crushed the cane of "outside" growers. The Natal Central Sugar Company was exceptional in that it was founded, in 1878, specifically to build a factory that would rely almost entirely on the cane of "outside" growers. The larger "miller-cum-planters" consolidated their activities, and after 1890 many incorporated, selling shares to the public to finance new developments. Three of the companies were incorporated in Britain, many of their shares were held there, and prominent among the shareholders were manufacturers of machinery for sugar factories. Through interlocking directorships and purchase of shares, management and ownership continued to coalesce and by 1914 one individual, Sir Charles Smith, had emerged as the dominant figure in the Natal industry (Osborn 1964: 177–327; Richardson 1984: 254–6).

Fiji provides an example of a vertically integrated company present from the beginning of a sugar industry. The Colonial Sugar Refining Company Ltd., of Australia, had been considering investment in Fiji since at least 1874, but despite the strong encouragement of the colonial government it delayed taking any initiative until 1880 when it built it first mill. Concerns about labor, landownership and competition for these resources from other crops all had to be allayed before the Company would commit itself. It was unwilling to depend entirely on Fijian labor to grow the cane, but Indian immigration soon disposed of this problem. In 1874, the future form of landownership in Fiji was still uncertain as the government tried to evolve a policy that would protect the rights of the native Fijians who had once controlled according to their own customs all the land in the islands, adjudicate the claims of European settlers to tracts they had alienated in one way or another from the Fijians and also ensure that there would be sufficient freehold land to support a plantation economy. During the late 1870s, these complicated matters were largely sorted out, many settlers confirmed in their ownership, and the government was able to offer the Company 1,000 acres at £2 an acre on the Rewa River in Viti Levu, with an option for 1,000 more on Vanua Levu, the main islands of the Fijian group. The final concern – fear of competition for land and labor – too disappeared as experiments with cotton, tea and coffee failed, and the Company realized the small sugar mills already on Fiji were unlikely to create problems for large modern mills of the type it would build. In 1880, it bought the first 1,000 acres and built a factory; it opened a second factory in 1883 and three more in the next ten years.

The mills of the Colonial Sugar Refining Company had several advantages over the competing mills. The Company concentrated its large financial resources on its basic activity, the sugar industry: it could afford to, and did, employ expert managers, chemists and technicians; it could spread the costs of research over its mills not only in Fiji but also in Australia; it could buy supplies in bulk; and it could transport the raw sugar in its own ships to its own refineries in Australia and New Zealand. These advantages of money, scale and vertical integration saw the Company's Fijian mills through the difficult years of the 1890s when prices were low. Only two competitors survived into the twentieth century, the Vancouver-Fiji Sugar Company and the Melbourne Trust Company, both of which owned a mill. When the Colonial Sugar Refining Company bought these mills in 1923 and 1926 respectively, it gained control of the entire Fijian industry (Lowndes 1956: 67–90; Moynagh 1981: 22–34).

The distinctive features of the financial organization of the sugar industry of Hawaii was the dominant role played by factors or agents. They were based in Honolulu and had their origins in the merchant houses that supplied the whaling ships during the first half of the nineteenth century. With the decline of whaling, these merchants had looked for other lines of activity, one of which was the sugar industry. They advanced cash to planters against the security of the crop, sold them imported tools and machinery, bought, warehoused, marketed and shipped the sugar they produced. The agencies made money out of the interest payments, profits on sales and fees for services, and gained a good deal of control over the planters (Adler 1966: 12–13, 92). The expansion of the industry after 1876 enormously benefited the agency business, but the pressures of competition led to amalgamations so that by the 1880s there were only seven important agencies left, a number that was eventually to be reduced to the "Big Five" of recent times (Adler 1966: 12–13, 92; Tate 1968: 122).

The existence of the agencies, however, did not prevent the development of a spectacular example of vertical integration that in its turn became a model for the Hawaiian industry. The entreprenuer was Claus Spreckels, a German immigrant to the United States, who had begun his career as an assistant in a South Carolina grocery store, but soon moved to California where his own grocery business flourished. He bought a brewery and in 1863 started the Bay Sugar Refinery. The sugar business clearly appealed to him, and, wanting to learn more about it, he sold his California holdings at great profit, and returned home to work in a sugar refinery in Magdeburg. In 1867, he was back in California where he immediately founded the California Sugar Refining Company. Spreckels at first opposed reciprocity with Hawaii fearing that the inflow of Hawaiian sugar would interfere with his plans for a beet sugar industry in California, but once the treaty was passed he determined to make the most of the new situation. He already knew something of the Hawaiian scene through his son John who in 1874 had gone to work there for H. Hackfeld and Company, a prominent German company which owned plantations and was also a suger agency; and in 1876 he visited Hawaii to assess the possibilities of making profits in sugar planting. He returned in 1878 to acquire cheaply large acreages on Maui in a series of controversial deals that enmeshed him in the political life of the islands. However, a prominent Californian irrigation engineer soon turned these barren lands into flourishing fields of cane. Spreckelsville plantation rapidly became extremely profitable and the main holding of the Hawaiian Commercial and Sugar Company in which initially Claus Spreckels was the majority stockholder. His next move was to acquire an agency so as to control the sale of his own sugar as well as to make profits from the sugar cane grown by other planters. In 1880, he decided to go into business with William G. Irwin, an Englishman, who had many business contacts in the islands and whose sugar agency, W. G. Irwin and Co., represented in Honolulu the firm of J. D. Spreckels and Brothers which John Spreckels had set up to import sugar for the California Sugar Refinery. The new partnership, carrying on business under the old name, soon became the most important agency, handling about half the Hawaiian crop in 1892. The final link in the integration was clasped with the founding of the Oceanic Steamship Line in 1881. All this had not been achieved without raising the envy and resentment of rival planters and agencies, and was not to last for long. He became out of touch with the politics of the islands, losing much of his prestige through his maneuverings against annexation. In 1893, he left Hawaii to look after his many interests in California; in 1898, his contacts with Hawaii were further weakened when his family lost control of the Hawaiian Commercial and Sugar Company. The contribution of individuals to economic change is often difficult to assess and Spreckels is still somewhat of a controversial figure, but without his energy, intelligence and money, probably the modernization of the Hawaiian sugar would not have gone ahead quite so rapidly (Adler 1966; Tate 1968: 121–30).

Other aspects of modernization of the sugar industry, familiar from the West Indies and elsewhere, were also present in these new colonies by 1914. Narrow-gauge railways were gradually being introduced. In Hawaii, the Spreckelsville plantation in 1881 was the first to lay tracks to bring cane from the fields to the mill. Some of the tracks were permament, others were designed to be moved around the plantation as harvesting progressed. For a few months, horses pulled the carts, but some locomotives were in service by late 1881 (Adler 1966: 75). Even on Réunion, the least progressive of these colonies, the larger estates by 1905 had systems of narrow-gauge railways built by the French firm of Decauville, although oxen rather than locomotives pulled the carts (Colson 1905: 340). Planters became involved in research into improving the cultivation and milling of cane. In Hawaii, Spreckels had given an enormous push in the direction of new technology and research, introducing at Spreckelsville, in addition to the railways, steam plows, five-roller mills, controlled irrigation, experimentation with fertilizers among other innovations (Adler 1966: 73–9). But research was put on a much more systematic basis by the Hawaiian Sugar Planters' Association. Founded in 1895 by the planters of Hawaii, and funded by them to look after their political and economic interests, it immediately established an agricultural research station that has over the years contributed enormously to success of the Hawaiian industry (Barnes 1974: 533–4; Beechert 1985: 179–80). The Colonial Sugar Refining Company has done much of its research in Australia, but it did establish a cane-breeding station in Fiji in 1904 to find replacements for cane being damaged by the so-called Fiji disease. The Fijian industry came to rely on varieties of cane either developed in Fiji or in Queensland (Barnes 1974: 325; Lowndes 1956: 75). On Mauritius a notable program of agricultural research began at a station founded in 1893 (Toussaint 1972: 247), but the island has a special place in the development of research in the industry for Mauritians were the first, according to Barnes (1974: 18), to employ chemists to control the work in sugar factories. The other two colonies were rather late to initiate research programmes – South Africa in 1925 and Réunion in 1928 – but only the South African Experiment Station became well-known for its work.

Australia

The sugar industry is located in the east of the country, in fertile river valleys along the coast from northern New South Wales through Queensland to Cairns. It dates from the 1860s, began to become of national importance during the last quarter of the nineteenth century and during recent decades has made Australia into a major exporter of sugar. The interest of this industry, and the reason it is given separate consideration here, derives not from its present size but from its distinctive characteristics. This is the only sugar cane industry in the world to rely on a labor force of European origin; it was also one of the first to rely on cane-farmers, either tenants or owners of small farms, to supply the central factories; and it pioneered mechanization of field work. The first two characteristics were already in place by 1914; mechanization of field work came later. This result could not have been predicted in the 1870s when the growing number of plantations with an imported, non-European work force presaged a standard pattern of development. The reorganization of the industry began in the years immediately following the collapse of the world price for sugar in 1885. The direction of change was in part influenced by technology and the economics of sugar production but it was also strongly influenced by the internal politics of Queensland and New South Wales and, after the federation of the Australian colonies in 1901, of Australia. In this democratic, self-governing country, the sugar industry in the early years of this century had to adapt to the type of society Australians wished to create.

The struggle, for such it was, between different interest groups centered on the type of labor in the industry and the manner in which the central factory system should be introduced. The two issues to an extent overlapped. The planters, as always and everywhere, wanted their labor supply to be as cheap and reliable as possible, but in Australia they had to attract men to the harsh labor of the cane fields in competition with other forms of agriculture and with mining. The government of Queensland, where the sugar lobby was powerful, accepted the argument, although not always easily and with a good many qualms, that imported indentured labor was necessary to keep down labor costs so as to make the industry viable. The Pacific Islands were the main source, and between 1863 and 1904, Queensland admitted 62,565 islanders or "Kanakas" (Saunders 1982: 97). The Queensland Polynesian Laborers Act of 1868 was an attempt to regulate a traffic which had many abuses: the laborers were to be engaged voluntarily for an indenture of three years after which they could either return home or re-engage but this time with the rights to choose their employer and bargain over wages and conditions of work. The "time-expired" laborers were also referred to as "overtime," "walking about" or "free" labor (Graves 1984: 266). New South Wales did not permit the entry of indentured laborers, but its planters were able to recruit "time-expired" Polynesians who came south from Queensland (Higman 1968: 704). A Queensland Act of 1884 introduced a third category of Pacific

Island worker, the so-called "ticket holders," who were immigrants admitted for unskilled work in "tropical industrial agriculture." This was a very small group, but its members earned their living in various ways, even as cane-farmers. Laborers also came, although in far smaller numbers, from India, Java, China, Ceylon and Japan. The planters attempted also to draw on the Aborigines. Despite the geographical range of this search, there had always been Australian laborers of British origin in the industry. As the European immigration to Australia increased, so more unskilled white laborers entered the industry, not only British, but Italian, Maltese and Portuguese. There was also a seasonal migration of labor from the Darling Downs during the slack season on the sheep runs and wheat farms to the cane fields (Graves 1984: 265, 271–2). With Europeans cutting cane, the belief sustained over the centuries that the white man could not do hard labor in the tropics was being shown to be a myth. This revelation, along with the growing participation of white workers, set the scene for an inter-racial competition for work.

The central factory system created yet other divisions. In 1868, when sugar production in New South Wales had barely begun, Edward Knox, the founder of the Colonial Sugar Refining Company, made a bid for a strong role in the industry and to secure supplies of sugar for his own refineries by placing a notice in a local newspaper to the effect that

IN RESPONSE TO NUMEROUS APPLICATIONS MADE TO THIS COMPANY TO ERECT SUGAR MILLS on the Northern Rivers of this colony, the DIRECTORS ... DESIRE TO NOTIFY that on being assured that a sufficient area of land has been planted to warrant the required outlay, they will be prepared to ESTABLISH CENTRAL SUGAR MILLS in the principal sugar growing districts.

The response was sufficiently encouraging because two mills were under construction in 1869 and they opened for business in 1870. The way into Queensland was made all the easier by the government which helped the Company to acquire land through the Colonial Sugar Refining Company Act of 1881 in return for which the Company agreed to invest £200,000 within five years (Birch and Blaxland 1956: 23, 27). The initial intention of the Company in New South Wales to rely on cane-farmers was not realized as the farmers showed, according to the Company, too much "independence," so it turned to settling tenants on its own land as well as to farming the land directly (Higman 1968: 702). In Queensland, too, the Company followed a policy of leasing its land to cane-farmers (Saunders 1982: 149). Australian planters, as did planters elsewhere in the world, at first resisted the new system, fearing loss of status if they gave up their mills and fearing also that they would become dependent on central factories which would dictate the price of cane just as now they dictated prices to their dependent cane-farmers. During the crisis years of the 1880s, the cane-farmers became more numerous as large holdings were subdivided and as an interest group they acquired a stronger political voice. They saw cooperative central factories, owned and run by themselves, as a means of improving their lot, and they petitioned the government of Queensland for assistance in setting them up. The Liberal government in 1885 acquiesced for political and economic reasons: it knew that the revival of the industry depended on the introduction of modern technology, but it wanted that revival to be based on a landowning democracy, on a society of cane-farmers who would support the Liberal Party rather than on large estates whose owners would vote for the Conservatives. After 1885, the central factories rapidly displaced the old mills both in Queensland and New South Wales (Graves 1984: 240, 276–7; Higman 1968: 710–11, 713; Saunders 1982: 144–53). By 1901, the date of federation, these developments had produced an industry with one powerful company, the Colonial Sugar Refining Company, which owned central factories as well as having a virtual monopoly of sugar refining in Australia, a number of privately owned as well as cooperatively owned central factories, a pattern of land tenure that contained not only large estates but also some thousands of tenant and freehold cane-farmers, a good many of whom worked their land with only their own and family labor and who thus represented a large addition to the white component in the labor force of the industry, and, at the bottom of the hierarchy, various categories of non-white imported laborers. The class structure of the industry had become more complex than before, and socio-economic tensions among the classes coexisted with the racial tensions within the labor force.

Australian nationalism incorporated a powerful component of racial intolerance as it gathered strength during the second half of the nineteenth century: Australia was to be a "British" dominion. An early manifestation of this intolerance occurred in the 1850s, in the gold fields of New South Wales and Victoria, when resentment over the success of Chinese miners led to legislation in both colonies to reserve mining for white labor; fifty years later the demand for a white Australia policy informed the negotiations that led to the federation of the colonies. Queensland would have to accept the national consensus on this matter. Among the very first Acts of the new Commonwealth parliament in 1901 were the Immigration Restriction Act and Pacific Islander Laborers Act which ordered the deportation of the islanders by 1907. The deportation was to be rigorously enforced, although in a "humanitarian" manner. Perhaps in order to mitigate this ambiguity, the government in 1906, shortly before the deadline, discovered some few compassionate grounds for granting exemptions to deportation, including in the reprieve those who had resided more than twenty years in the country, and/or owned land or held unexpired leaseholds, but true to the intent of the Acts by 1907 the great majority of the Pacific islanders had left Australia. In return for agreeing to become part of white Australia, Queensland was given free access for its sugar to the Australian market which in turn would be protected by tariffs against foreign cane and beet sugar. Additionally, to alleviate any extra cost the change in labor force might cause, as well as to counter the charge that it was taking away Pacific island labor to replace it with local Chinese, the Commonwealth government agreed in 1902 to pay for an initial period of five years a bonus of £3 a ton to cane-growers who employed only white labor.

As white Australians of British descent moved into the cane fields, the Australian Federation of Labor followed. However, the Sugar Workers' Union, later incorporated into the Amalgamated Workers' Association, was founded in 1906 to look after the interests of this one group of workers, excluding from membership the Chinese, Japanese and few remaining islanders, and although the Italians who were now beginning to enter Queensland were, as Europeans, eligible to join they were not encouraged to do so. One consequence of this racism was that the Union was unable to present a united front of workers in the bitter strike of 1911; spurned and vulnerable, the Italians and non-whites continued to work – strike-breakers in the eyes of Union members – for the Colonial Sugar Refining Company and other wealthy employers, which of course further exacerbated relations between Union members, non-members and the employers. After the strike, both government and Union continued to pursue the "white Australia" policy. The 1913 Sugar Works Guarantee Act, which was aimed at the few remaining islanders in the industry, forbade aliens from cultivating cane on their own land and from leasing land to cultivate cane. The Union made plain to farmers that life might become difficult for them if they employed non-white labor. Chinese and Japanese moved on elsewhere into other lines of activity; the last islanders, too, left the industry they had helped build to live out their lives unwanted and in poverty on the margins of Australian society (Saunders 1982: 144–83).

Depending on one's perspective, the Colonial Sugar Refining Company played a pioneering or controversial role in the industry, progressive in its introduction of cane-farming and the central factory system, but on the other hand employing non-union labor and enjoying a near-monopoly of sugar refining in Australia. Less controversial, certainly, is the Company's record in support of research. As early as 1870, it published a pamphlet of "Plain Hints to Plain Men for the Cultivation of Sugar Cane" and since 1891 has followed a policy of cane breeding on its own experimental stations. It has also supported the Queensland Bureau of Sugar Experiment Stations which was founded in 1901 (Dixon 1956: 124–30). Queensland has come to rely very heavily on varieties of cane produced by the Company. The early introduction of railways and modern mills has helped the efficiency

The Indian Ocean and Pacific colonies: 1800–1914   233

of the industry, and while experiments to mechanize cane harvesting began in the early years of this century, little was achieved until well after 1914 (Willis 1972: 11). A measure of the success of the industry lies in the production figures: in 1870, Queensland produced 2,900 tons of raw sugar and in 1907, 191,320 tons (Graves 1984: 264).

With the creation of these sugar colonies in the Pacific, sugar cane had in a literal sense come full-circle: domesticated in New Guinea it spread north and west and finally returned to the Pacific, much changed, as an industrial crop. In the new colonies of the Pacific, as centuries before around the Mediterranean, on the Atlantic islands and in colonial America, the demands of the sugar industry transformed landscapes, destroyed indigenous societies, triggered flows of immigrants and created a world of its own. It was, along with the western imperialism with which it was so closely linked, a major factor in the formation of the cultural geography of the tropical world. The date, 1914, as in so many aspects of history, represents in retrospect a convenient break point after which the evolution of the sugar cane industry took on a rather different pattern in response to new political, technological and economic pressures.

Conclusion

The geography of sugar production at the outbreak of the First World War can be easily outlined. It had three important components: centrifugal sugar from cane, non-centrifugal sugar from cane and sugar from beet. World production of centrifugal cane sugar in 1914 amounted to some 8,000,000 tons. Although there were major producers with their own domestic markets, such as Hawaii, Louisiana, Natal, Queensland, Argentina and Brazil, most of the centrifugal sugar entered international trade which flowed to North America, Europe, India, China and, increasingly, to Japan. Cuba alone exported nearly 2,600,000 tons; Java with exports of about 1,350,000 tons was its nearest rival. The weight of production was still in the Americas which were responsible for perhaps 5,000,000 tons. However, with a combined production of about 3,000,000 million tons, the new participants in the Indian Ocean, southeast Asia and the Pacific had long since broken the near-monopoly on the cane sugar trade that the Americas had once held and they were indeed gaining in importance. Statistics on the production of non-centrifugal cane sugar are probably little better than informed guesses. Only small quantities of non-centrifugal sugar entered international trade and are thus traceable; much of the production of *gur, rapadura, panela, pilon* and other types of muscovado sugar made in many thousands of small mills for local markets in the hinterlands of the tropical world went unreported to any government authority. It is reasonable to conclude, given the numbers of traditional mills still in operation and the slowness with which the central factory system was being introduced, that in China and India, and probably also in Brazil and Colombia, the production of non-centrifugal sugar exceeded that of centrifugal. According to one estimate (Lowndes 1956: 440), the total production of all cane sugar in 1914 was 10,000,000 tons, and as 8,000,000 of this amount was centrifugal, there were presumably 2,000,000 tons of non-centrifugal sugar made in that year but it is not easy to place great confidence in this figure. Beet sugar before the First World War was still largely a

European industry: of the 6,581,000 tons produced in 1910, Europe accounted for 6,074,000 and the United States for 507,000. Total beet production reached a prewar peak of just over 9,000,000 tons in 1913. Sugar beet was grown in most of the countries of continental Europe from Spain across to Russia and up to Sweden. The crop was particularly important in northern France, in Germany, especially around Magdeburg, in Bohemia and the Ukraine. Germany was the main producer; Germany and Austria–Hungary were the major exporters; Great Britain was the major importer. Great Britain did not have a sugar beet industry, preferring to benefit instead from the subsidies of other governments rather than subsidize an industry of its own. In 1913, beet sugar made up 87% of British sugar imports, the greater part supplied by the two continental empires with which it was about to go to war (Chalmin 1983: 14–28; 1984: 15).

By 1914, cane sugar had made a good recovery from the initial competition from beet. Comparisons of the contribution of beet and cane sugar to total sugar production probably underestimate the contribution of cane because of the uncertainty over the amount of non-centrifugal sugar in the world, and the following figures from Lowndes (1956: 438–41) are offered with this proviso. The 1890s were the nadir for cane when its share of the total dropped below 40%, but a recovery followed in the early years of the century and was up again over 50% by 1914. A good part of the War was fought in fields of sugar beet and this devastation is reflected in the statistics: in 1919, cane contributed 78% of world sugar production, a percentage it has not subsequently surpassed.

The sugar cane industry has continued to evolve since 1914 in response to economic and political forces, and in the process has continued to be a force of landscape change. The First World War was followed by an era in which large multinational companies dominated much of the industry; after the Second World War many an ex-colonial now independent country nationalized the industry and tried to make it more "democratic" by promoting cane-farming. Incessant competition from beet, joined more recently by high fructose corn syrup and artificial sweeteners, has forced cane producers to strive for ever greater efficiency in the fields and factories, and to search for new, different markets. Bagasse is now a commercial fuel and raw material for the manufacture of heavy duty board and card; alcohol from cane drives cars in Brazil; immature cane is fed to cattle; and other uses for the products of sugar cane no doubt will be found. Despite these successes, symptomatic of the sugar cane industry's ability to respond to competition, some of the smaller producers have gone or may soon go out of the business. In addition to competition, an additional problem for such producers has been the increase in the optimum size of central factories with the consequence that they cannot grow sufficient cane to feed a modern factory. As Nevis bowed out earlier in this century, so now Antigua is withdrawing from the industry which has been the mainstay of its economy for three centuries. France has opted to protect beet producers at home rather than cane producers overseas and so Martinique, Guadeloupe and Réunion are slipping out of the industry. St. Kitts and Barbados may follow in the not too distant future. The final chapter in the historical geography of the industry must be an analysis of the problems of adjusting to a post-sugar economy. The competitiveness of the world sugar market together with the huge vested interests at stake, both those of the producers – whether of cane, beet or high fructose corn syrup – and of the consumers, has ensured that the sugar industry continues to be a managed industry, with an ample diplomatic work force negotiating and re-negotiating prices, quotas, livelihoods. In the late twentieth century the geography of the sugar cane industry is a fascinating result of the variation in climate and soils around the tropical world, of the lobbying of the beet producers in the E.E.C., of the lobbying of beet, corn and cane producers in the United States, of Soviet policy towards Cuba and even of the sense of compassion and/or of guilt of ex-mother countries for the ex-tropical colonies that leads to guarantees to purchase given quantities of cane sugar at given prices.

Mankind, however, has retained his taste for sweetness, and while the few very rich societies are beginning to reduce their consumption of sugar that correlation between increasing wealth and increasing sugar consumption, so noticeable in England during the eighteenth and nineteenth centuries, is still holding true today in the developing world. There will surely be an increase in the world consumption of sweeteners, and if the Chinese delight in soft drinks grows, as there is some suggestion it might among the younger generation, the upward curve in world demand may prove to be very steep indeed.

The pre-1914, pre-central factory landscape of the sugar cane industry is now submerged by later accretions of cultural and economic life, whether yet bigger central factories, suburban development, cotton fields, vineyards or tourist villas, but it has not gone; it can still be seen, read and appreciated by those with the knowledge in many parts of the Mediterranean and the tropics. Archaeologists are unearthing mills and irrigation systems in Cyprus, sugar cane still grows in a few small fields on the terraces of Madeira. At the end of many a red rutted-clay road in the Zona da Mata of Pernambuco there are collections of buildings many now ruinate: mansion, factory, chimney, chapel. Sometimes the avenue of royal palms leads only to an open space, more cane fields, a place name and a site that has meaning for the descendants of those who once worked there, both master and slave. In the West Indies, plantation houses have become hotels for the more discriminating vacationers, though one wonders whether they rest easily if they know how and why their hotels were built, the surrounding landscape formed. The symbolism in this landscape is indeed a difficult inheritance, and preservation requires a continuing commitment to honesty: to restore not only the baroque churches and Georgian great houses but also the slave markets as has been done in Olinda, Pernambuco. And the people whom this cruel, once-prosperous industry brought together are still there, in those green fields behind coral beaches in lands and islands with lilting, lovely, evocative names.

Appendix: the price of sugar

The earliest statements on the price of sugar in the western world occur in the accounts of merchants and in the pantry records of noble households. Later, from the 1600s, information on the price of sugar in the main commercial centers of Europe is abundant, and so it is possible to graph on a year by year basis the downward trend in prices, a trend marked by occasional short-lived upward oscillations due to war or other crises. It is, however, less easy to correct the prices for inflation, and the downward trend was in fact far steeper than the simple plotting of the record suggests. Figures A.1 and A.2 make no allowance for inflation. Figure A.1, first published by Lyle (1950) is perhaps the most ambitious attempt to retrace the movement of the price of sugar. The graph shows clearly the depressing influence on price of the arrival in Europe of sugar from the Atlantic islands after 1450, but the great inflation in Europe that followed the import of American gold and silver obscures the impact of the early Brazilian industry. The second half of the graph records two extended periods of a major decline in price: following 1650 when the new English and French Caribbean colonies began to export sugar and following 1850 when competition from beet became intense. The wars of the second half of the eighteenth century (the Seven Years' War, the War of American Independence), the Napoleonic Wars and the First World War produced an increase in price. Figure A.2 shows in greater detail the movement in price between 1730 and 1914. Lyle (1950: 533) tried to estimate the decline over the centuries in the purchasing power represented by the price of 1 pound of sugar (Fig. A.1). By way of illustration: between 1259 and 1400, 1 pound of sugar was equal in price to 28 pounds of cheese, or 29 pounds of butter, or thirty-four dozen eggs; in 1937 the price of 1 pound of sugar would buy only 3 ounces of cheese, or 3 ounces of butter or two eggs.

The price of sugar 239

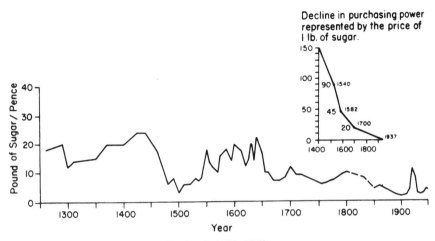

Fig. A.1 Retail price of sugar, England, 1259–1950
Source: after Lyle 1950: 532–3.

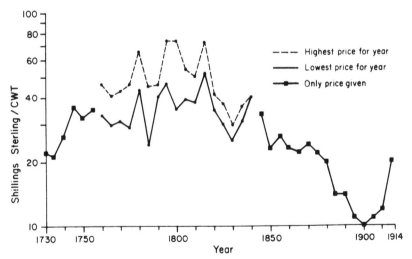

Fig. A.2 Raw sugar (cost, insurance, freight) in London, 1730–1914 (five-year averages)
Source: Lowndes 1956: 442–3.

List of references

Adamson, Alan H. 1972. *Sugar without slaves. The political economy of British Guiana, 1838–1904*. New Haven and London, Yale University Press
Adler, Jacob. 1966. *Claus Spreckels, the sugar king in Hawaii*. Honolulu, University of Hawaii Press
Ajuda Palace Library, Lisbon. 51–V1–54, folio 121
Albert, Bill. 1976. *An essay on the Peruvian sugar industry 1880–1920 and the letters of Ronald Gordon, administrator of the British Sugar Company in Cañete, 1914–1920*. Norwich, School of Social Studies, University of East Anglia
  1984. The labour force on Peru's sugar plantations 1820–1930: a survey. In Albert and Graves, pp. 199–215
Albert, Bill, and Graves, Adrian (eds.). 1984. *Crisis and change in the international sugar economy 1860–1914*. Norwich and Edinburgh, I.S.C. Press
Alden, Dauril. 1963. The population of Brazil in the late eighteenth century: a preliminary study. *Hispanic American Historical Review*, **43**, 173–205
  1984. Late colonial Brazil, 1750–1808. In Leslie Bethell (ed.), *The Cambridge history of Latin America*, vol. 2, pp. 601–60. Cambridge, Cambridge University Press
Allen, G. C., and Donnithorne, Audrey G. 1962. *Western enterprise in Far Eastern economic development. China and Japan*. London, George Allen and Unwin
Amin, Shahid. 1984. *Sugarcane and sugar in Gorakhpur. An inquiry into peasant production for capitalist enterprise in colonial India*. Delhi, Oxford University Press
Anderson, E. N. Jr., and Anderson, Marja L. 1977. Modern China: South. In K. C. Chang (ed.), *Food in Chinese culture. Anthropological and historical perspectives*, pp. 317–82. New Haven and London, Yale University Press
Anonymous. 1956. *Diálogos das grandezas do Brasil*. Salvador, Livraria Progresso Editôra (The MS of this work dates from 1618.)
Antonil, André João. 1968. *Cultura e opulência do Brazil por suas drogas e minas*. Translation and critical commentary by André Mansuy, Paris, Institut des Hautes Etudes de L'Amérique Latine
Arango y Parreño, Francisco. 1952. *Obras de D. Francisco de Arango y Parreño*. 2 vols., La Habana, Dirección de Cultura, Ministerio de Educación

Ashtor, Eliyahu. 1969. *Histoire des prix et des salaires dans l'Orient médiévale*. Paris, S.E.V.P.E.N.

1981. Levantine sugar industry in the late Middle Ages, a sample of technological decline. In A. L. Udovitch (ed.), *The Islamic Middle East, 700–1900*, pp. 91–132. Princeton, Darwin Press

Avalle, Le Citoyen. 1799. *Tableau comparatif des colonies françaises aux Antilles, avec celles des colonies anglaises, espagnoles, et hollandaises, de l'année 1787 à 1788. Suivi de l'établissement et mouvement d'une sucrerie, pendant le cours d'une année*. Paris, Chez Goujons fils, Debray, Fuschs, An VII, 27 Fructidor

Bailey, L. H. 1961. *Manual of cultivated plants*. New York, Macmillan

Banerjee, Himadri. 1982. *The agrarian society of the Punjab, 1849–1901*. New Delhi, Manohar

Banerji, Sures Chandra. 1971. *A companion to Sanskrit literature*. Delhi, Varanasi, Patna, Motilal Banarsidass

Barnes, A. C. 1974. *The sugar cane*. Aylesbury, Leonard Hill Books

Barrett, Ward. 1965. Caribbean sugar production standards in the seventeenth and eighteenth centuries. In John Parker (ed.), *Merchants and scholars. Essays in the history of exploration and trade*, pp. 147–70. Minneapolis, University of Minnesota Press

1970. *The sugar hacienda of the marquesas del Valle*. Minneapolis, University of Minnesota Press

1976. Morelos and its sugar industry in the late eighteenth century. In Ida Altman and James Lockhart (eds.), *Provinces of early Mexico*, pp. 155–75. Latin American Center Publications, no. 1, Los Angeles, University of California

1979. *The efficient plantation and the inefficient hacienda*. The James Ford Bell Lectures, no. 16, Minneapolis, University of Minnesota Press

Barrett, Ward, and Schwartz, Stuart B. 1975. Comparación entre dos economias azucareiras coloniales: Morelos, México y Bahia, Brasil. In Enrique Florescano (ed.), *Haciendas, latifundias y plantaciónes en America Latina*, pp. 532–72. México, Madrid, Buenos Aires, Siglo Veintiuno Editores

Baxa, Jacob, and Bruhns, Guntwin. 1967. *Zucker im Leben der Völker. Eine Kultur- und Wirtschaftsgeschichte*. Berlin, Verlag Dr. Albert Bartens

Beachey, R. W. 1957. *The British West Indies sugar industry in the late 19th century*. Oxford, Basil Blackwell

Beckford, William. 1790. *A descriptive account of the island of Jamaica: with remarks on the cultivation of sugarcane . . . also observations and reflections upon what would probably be the consequences of an abolition of the slave trade, and of the emancipation of the slaves*. 2 vols., London, T. and J. Egerton

Beechert, Edward D. 1985. *Working in Hawaii*. Honolulu, University of Hawaii Press

Bell, Roger. 1984. *Last among equals. Hawaiian statehood and American politics*. Honolulu, University of Hawaii Press

Benvenisti, Meron. 1970. *The Crusaders in the Holy Land*. Jerusalem, Israel University Press

Bergad, Laird W. 1978. Agrarian history of Puerto Rico, 1870–1930. *Latin American Research Review*, **13**, 63–94

Berthier, Paul. 1966a. *Les Anciennes Sucreries du Maroc et leurs réseaux hydrauliques. Etude archéologique et d'histoire économique. Un épisode de l'histoire de la canne à sucre.* 2 vols., Rabat, Imprimeries Française et Marocaines
  1966b. Les Plantations de canna à sucre et les fabriques de sucre dans l'ancien Maroc. *Hespéris-Tamuda*, 7, 33–40
Bibliothèque Nationale, Paris, Dept. des Cartes et des Plans
  a) Martinique: S. H. Portefeuille 156, Div. 2, Pièces 9.B and 12.D; Guadeloupe: S. H. Portefeuille 155, Div. 2, Pièce 4.B
  b) St. Domingue: Portefeuille 150, Div. 4, Pièce 11.D
Birch, Alan, and Blaxland, A. F. 1956. The historical background. In A. G. Lowndes (ed.), *South Pacific enterprise. The Colonial Sugar Refining Company Limited*, pp. 11–53. Sydney, Angus and Robertson
Blume, Helmut. 1958. El cultivo de la caña de azúcar en Andalucia comparado con el cultivo de la caña en Louisiana. *Estudios Geográficos*, 19, 87–112
  1985. *Geography of sugar cane.* Berlin, Verlag Dr. Albert Bartens
Bolens, Lucie. 1972a. Engrais et protection de la fertilité dans l'agronomie hispano-avabes XIe–XIIe siècles. *Etudes Rurales*, 46, 34–60
  1972b. L'Eau et l'irrigation d'après les traités d'agronomie andalous au moyen âge (XIIe siècle). *Options Méditerranéenes*, 16, 65–77
Bowser, Frederick P. 1984. Africans in Spanish American colonial society. In Leslie Bethell (ed.), *The Cambridge history of Latin America*, vol. 2, pp. 357–9. Cambridge, Cambridge University Press
Boxer, C. R. 1969. *The Portuguese seaborne empire 1415–1825.* London, Hutchinson
Braudel, Fernand. 1972–3. *The Mediterranean and the Mediterranean world in the age of Phillip II.* Trans. Siân Reynolds, 2 vols., London, Collins
  1982. *The wheels of commerce*, vol. 2, in *Civilization and capitalism, 15th–18th century.* Trans. Siân Reynolds, London, Collins
Bray, Francesca. 1984. *Agriculture*, vol. 6, part 2, in Joseph Needham, *Science and civilization in China.* Cambridge, Cambridge University Press
  1986. *The rice economies. Technology and development in Asian societies.* Oxford, Basil Blackwell
Brockway, Lucille, H. 1979. *Science and colonial expansion. The role of the British Royal Botanic Gardens.* New York and London, Academic Press
Brothwell, Don, and Brothwell, Patricia. 1969. *Food in antiquity. A survey of the diet of early peoples.* New York and Washington, Frederick A. Praeger
Bry, Theodor de. 1595. *Collectiones peregrinatiorum. Americas pars quinto hia Hispanorum*, Francofurti ad Moenum, n.p.
Bryan, Patrick. 1978. The transition of plantation agriculture in the Dominican Republic, 1870–84. *Journal of Caribbean History*, 10/11, 82–105
Buchanan, Francis. 1807. *Journey from Madras through the countries of Mysore, Canara and Malabar performed under the orders of the most noble Marquis of Wellesley, Governor General of India, for the express purpose of investigating the state of agriculture, arts, and commerce; the religion, manners, and customs; the history natural and civil, and antiquities in the dominions of the Rajah of Mysore, and the countries acquired by the Honourable East India Company, in the late and former wars, from Tippoo Sultan.* 3 vols., London, T. Cadell

and W. Davies, and Black, Parry and Kingsbury
Burnett, John. 1966. *Plenty and want. A social history of diet in England from 1815 to the present day*. Edinburgh, Nelson
Burns, E. Bradford. 1980. *The poverty of progress. Latin America in the nineteenth century*. Berkeley, University of California Press
Cahen, Claude. 1940. *La Syrie du nord à l'époque des Croissades et la Principauté Franque d'Antioche*. Paris, Librairie Orientaliste Paul Geuthner
  1968. *Pre-Ottoman Turkey. A general survey of the material and spiritual culture and history 1071–1330*. London, Sidgwick and Jackson
Camacho, Guillermo, and Galdas, Perez. 1961. El cultivo de la caña de azucar y la industria azucarero en Gran Canaria, 1510–1535. *Anuario de Estudios Atlanticos*, 7, 11–60
Canabrava, Alice Piffer. 1981. *O açúcar nas antilhas (1697–1755)*. São Paulo, Instituto de Pesquisas Econômicos
Carli, Gileno dé. 1943. *Geñnese e evolução da indústria a ucareira de São Paulo*, Rio de Janeiro, Pongetti
Cassá, Roberto. 1979–80. *Historia social y económica de la Republica Dominicana*. 2 vols., Santo Domingo, Editora Alfa y Omega
Castillo, José del, and Cordero, Walter 1980. *La economia Dominicana durante el primer cuarto del siglo XX*, Santo Domingo, Fundación García-Arévalo
Chalmin, Philippe. 1983. *Tate and Lyle. Géant du Sucre*. Paris, Economica
  1984. The important trends in sugar diplomacy before 1914. In Albert and Graves, pp. 9–19
Chapoutet-Remadi, Mounira. 1974. L'Agriculture dans l'Empire Mamluk au moyen âge. *Les Cahiers de Tunisie*, **85–6**, 23–45
Chaudhuri, B. 1983. Regional economy (1757–1857): Eastern India (2). In Dharma Kumar (ed.), with the editorial assistance of Meghdad Desai, *The Cambridge economic history of India*, vol. 2, c. 1757–c. 1970, pp. 295–332. Cambridge, Cambridge University Press.
Chaudhuri, K. N. (ed.). 1971. *The economic development of India under The East India Company 1814–1858*. Cambridge, Cambridge University Press
  1985. *Trade and civilization in the Indian Ocean: an economic history from the rise of Islam to 1750*. London and New York, Cambridge University Press
Chaunu, Huquette, and Chaunu, Pierre. 1955–6. *Séville et l'Atlantique (1504–1650)*. 5 vols. (1–5), Paris, Librairie Armand Colin.
  1956–9. *Séville et l'Atlantique (1504–1650)*. 3 vols. (6–8), Paris, S.E.V.P.E.N.
Chélus, A. de. 1719. *Histoire naturelle du cacao et du sucre*. Paris, Chez Laurant d'Houry
Chen, James C. P. 1985. *Cane sugar handbook*. New York, John Wiley and Sons
Clark, W. 1823. *Ten views of the island of Antigua*. London, Thomas Clay
Clarke, Colin G. 1975. *Kingston, Jamaica. Urban growth and social change, 1692–1962*. Berkeley, University of California Press
Clifton, James N. 1981. The rice industry in colonial America. *Agricultural History*, **55**, 266–83
Colson, Léon. 1905. *Culture et industrie de la canne à sucre aux îles Hawai et à la Réunion*. Paris, Augustin Challamel

Cook, Sherburne F., and Borah, Woodrow. 1971. *Essays in population history. Mexico and the Caribbean*, vol. 1. Berkeley, Los Angeles and London, University of California Press

Corwin, Arthur F. 1967. *Spain and the abolition of slavery in Cuba, 1817–1886.* Austin and London, University of Texas Press

Coulter, Ellis Merton. 1940. *Thomas Spalding of Sapelo.* University, La., Louisiana State University Press

Cramer, P. J. S. 1952. Sugar-cane breeding in Java. *Economic Botany*, **6**, 143–50

Cunha, Antônio Geraldo da. 1982. *Dicionário histórico das palavras Portuguesas de origem Tupi.* São Paulo, Melhoramentos

Curtin, Philip D. 1969. *The Atlantic slave trade. A census.* Madison, University of Wisconsin Press

Cushner, Nicholas P. 1980. *Lords of the land. Sugar, wine and Jesuit estates of coastal Peru, 1600–1767.* Albany, State University of New York Press

Dalechamps, Jacques. 1615. *Histoire générale des plantes.* 2 vols., trans. Jean des Moulins, Lyons, Chez les Heritiers Guillaume Rouille

Darrag, Ahmad. 1961. *L'Egypt sous le règne de Barsbay 825–841/1422–1438.* Damas, Institut Français de Damas

Daubrée, Paul. 1841. *La Question coloniale sous le rapport industriel.* Paris, F. Malteste

David, Elizabeth. 1977. *English bread and yeast cookery.* London, Allen Lane, Penguin Books Ltd.

Davis, Ralph. 1973. *The rise of the Atlantic economies.* Ithaca, N.Y., Cornell University Press

Debien, Gabriel. 1962. *Plantations et esclaves à Saint Domingue.* Dakar, Université de Dakar, Faculté des Lettres et Sciences Humaines, Publications de la Section d'Histoire, no. 3

Deerr, Noel. 1949–50. *The history of sugar.* 2 vols., London, Chapman and Hall Ltd.

Dixon, J. M. 1956. Sugar milling by C.S.R. In A. G. Lowndes (ed.), *South Pacific enterprise. The Colonial Sugar Refining Company Limited*, pp. 119–45. Sydney, Angus and Robertson

Dols, Michael W. 1977. *The Black Death in the Middle Fast.* Princeton, Princeton University Press

Donkin, R. A. 1980. *Manna. An historical geography.* The Hague, W. Junk B. V. Publishers

Dozy, R., and Pellat, Ch. 1961. *Le Calendrier de Cordoue.* Leiden, E. J. Brill

Drummond, J. C., and Wilbraham, A. 1939. *The Englishman's food. A history of five centuries of English diet.* London, Jonathan Cape

Duncan, T. Bentley. 1972. *Atlantic islands. Madeira, the Azores and the Cape Verdes in seventeenth-century commerce and navigation.* Chicago and London, University of Chicago Press

Dunn, Richard S. 1969. The Barbados census of 1680: profile of the richest colony in English America. *William and Mary Quarterly*, 3rd ser., **26**, 3–30

 1972. *Sugar and slavery. The rise of the planter class in the English West Indies, 1624–1713.* Published for the Institute of Early American History and Culture

at Williamsburg, Virginia, Chapel Hill, University of North Carolina Press
Edel, Matthew. 1969. *The Brazilian sugar cycle of the seventeenth century and the rise of West Indian competition. Caribbean Studies*, **9**, 24–44
Edwards, Bryan. 1793–1801. *The history, civil and commercial of the British colonies in the West Indies*. 3 vols., London, John Stockdale
Eisenberg, Peter L. 1974. *The sugar industry in Pernambuco, 1840–1910. Modernization without change*. Berkeley, University of California Press
Eisner, Gisella. 1961. *Jamaica 1830–1930*. Manchester, University of Manchester Press
Elson, R. E. 1984. *Javanese peasants and the colonial sugar industry. Impact and change in an East Java residency 1830–1940*. Singapore, Oxford University Press
Emmanuel, Isaac S., and Suzanne, A. 1970. *History of the Jews of the Netherlands Antilles*. 2 vols., Cincinnati, American Jewish Archives
Fabrellas, Maria Luisa. 1952. La producción de azúcar en Tenerife. *Revista de Historia* (La Laguna), **18**, 455–80
Fagg, John Edwin. 1965. *Cuba, Haiti and the Dominican Republic*. Englewood Cliffs, Prentice-Hall
Feldhaus, F. M. 1954. *Die Machine im Leben der Volker*. Basel and Stuttgart, Verlag Birkhauser
Fernandez-Armesto, Felipe. 1982. *The Canary Islands after the conquest. The making of a colonial society in the early sixteenth century*. Oxford, Clarendon Press
Fraginals, Manuel Moreno. 1976. *The sugarmill. The socioeconomic complex of sugar in Cuba 1760–1860*. New York and London, Monthly Review Press
Freyre, Gilberto. 1933. *Casa-Grande e Senzala*. Rio de Janeiro, Maia e Schmidt. Trans. by Samuel Putnam. 1946. *The masters and the slaves*. New York, Alfred A. Knopf
Fry, James. 1987. Sweetener production, consumption and price cycles 1987–1990. The world picture. *Sugar y Azúcar*, April, 14–21
Furnas, J. C. 1969. *The Americans. A social history of the United States, 1587–1914*. New York, G. P. Putnam and Sons
Furnivall, J. S. 1939. *Netherlands India. A study of plural economy*. Cambridge, Cambridge University Press
Gage, Thomas. 1928. *The English-American. A new survey of the West Indies, 1648*. Ed. A. P. Newton, London, G. Routledge and Sons
Galloway, J. H. 1964. The sugar industry in Barbados during the seventeenth century. *Journal of Tropical Geography*, **19**, 35–41
  1968. The sugar industry of Pernambuco during the nineteenth century. *Annals of the Association of American Geographers*, **58**, 285–303
  1971. The last years of slavery on the sugar plantations of northeast Brazil. *Hispanic American Historical Review*, **51**, 586–605
  1977. The Mediterranean sugar industry. *Geographical Review*, **67**, 177–94
  1979. Agricultural reform and the Enlightenment in late colonial Brazil. *Agricultural History*, **53**, 763–9
  1985. Tradition and innovation in the American sugar industry, c. 1500–1800:

an explanation. *Annals of the Association of American Geographers*, **75**, 334–51
Gama, Rui. 1983. *Engenho e Tecnologia*. São Paulo, Livraria Duas Cidades
Garcia, Carlos Alberto. 1966. A ilha de S. Tomé como centro experimental do comportamento do Luso nos trópicos. *Studia*, **19**, 209–21
Garfield, Robert. 1971. *A history of São Tomé island, 1470–1655*. Ann Arbor, University Microfilms
Geertz, Clifford. 1968. *Agricultural involution. The process of ecological change in Indonesia*. Berkeley and Los Angeles, University of California Press
Ghee, Lim Teck. 1977. *Peasants and their agricultural economy in colonial Malaya 1874–1941*. Kuala Lumpur, Oxford University Press
Gil-Bermejo Garcia, Juana. 1970. *Panorama histórico de la agricultura de Puerto Rico*. Sevilla, Instituto de Cultura Puertoriqueña
Glasse, Hannah. 1760. *The compleat confectioner*. London, sold at Mrs. Ashburner's shop
Glick, Thomas F. 1972. *The old world background of the irrigation system of San Antonio, Texas*. Southwestern Studies, no. 35, El Paso, Texas Western Press
1979. *Islamic and Christian Spain in the early Middle Ages*. Princeton, Princeton University Press
Gonzales, Michael J. 1985. *Plantation agriculture and social control in Northern Peru, 1875–1933*. Austin, University of Texas Press
Gopal, Lallanji. 1964. Sugar-making in ancient India. *Journal of the Economic and Social History of the Orient*, **7**, 57–72
Grainger, James. 1764. *The sugar-cane: a poem*. London, R. and J. Dodsley
Graves, Adrian. 1984. Crisis and change in the Queensland sugar industry, 1862–1906. In Albert and Graves, pp. 256–79
Great Britain. 1880. *Calendar of state papers, colonial series, America and West Indies*, vol. 5. London, H.M. Stationary Office
Greenfield, Sidney M. 1977. Madeira and the beginnings of New World sugar cultivation and plantation slavery: a study in institution building. In Vera Rubin and Arthur Tuden (eds.), *Comparative perspectives in New World plantation societies*, pp. 536–52. The Annals of the New York Academy of Sciences, no. 293, New York, New York Academy of Sciences
Gregorio, Rosario. 1845. *Opere scelte degli zuccheri Siciliana*. Palermo, Tipografia di Pietro Pensante
Gupta, N. S. 1970. *Industrial structure of India during medieval period (sic.)*. Delhi, S. Chand and Co.
Guy, J. Donna. 1980. *Argentine sugar politics. Tucumán and the generation of eighty*. Tempe, Arizona State University
1984. Sugar industries at the periphery of the world market: Argentina 1860–1914. In Albert and Graves, pp. 147–62
Hall, Douglas. 1959. *Free Jamaica 1838–1856. An economic history*. New Haven, Yale University Press
1961. Incalculability as a feature of sugar production during the eighteenth century. *Social and Economic Studies*, **10**, 340–52
1971. *Five of the Leewards 1834–1870. The major problems of the post-emancipation period in Antigua, Barbuda, Montserrat, Nevis and St. Kitts*. Barbados,

Caribbean Universities Press

Haraksingh, Kusha. 1984. Technology and sugar estates in Trinidad, 1870–1914. In Albert and Graves, pp. 133–45

Heers, Jacques H. 1957. Le Royaume de Grenade et la politique marchande de gênes en occident (XV siècle). *Le Moyen Age*, **63**, 87–121

1961 *Gênes au XV siècle. Activité économique et problèmes sociaux*. Paris, S.E.V.-P.E.N.

Hernandez, Francisco. 1615. *Quatro libros de la naturaleza y virtudes de las plantas, y animales que estan receuidos en el uso de medicina en la Nueva Espana, y la methodo, y correcion, y preparacion, que para administrallas se requiere con lo que el Doctor Franciso Hernandez escrivio en lengua Latina. Traduzido y aumentados muchos simples, y compuestos y otros muchos secretos curativos, por Fr. Franciso Ximenez*. México: at the house of the widow of Diego Lopez Davalos

Hernandez, Ramon Diaz. 1982. *El azúcar en Canarias (S. XVI–XVII)*. Las Palmas de Gran Canaria, La Mancomunidad de Cabildos, Plan Cultural, y Museo Canario

Hess, John L., and Hess, Karen, 1977. *The Taste of America*. New York, Grossman Publishers

Heyd, Wilhelm von. 1879. *Geschichte des Levantehandels im Mittelalter*. 2 vols., Stuttgart, J. G. Cotta

Higman, B. W. 1968. Sugar plantations and yeoman farming in New South Wales. *Annals of the Association of American Geographers*, **4**, 697–719

Hill, Sir George Francis. 1948. *A history of Cyprus*. 3 vols., Cambridge, Cambridge University Press

Hilliard, Sam B. 1979. Site characteristics and spatial stability of the Louisiana sugarcane industry. *Agricultural History*, **53**, 254–69

Hirschmeier, Johannes. 1964. *The origins of entrepreneurship in Meiji Japan*. Cambridge, Mass., Harvard University Press

Hommel, Rudolf P. 1969. *China at work. An illustrated record of the primitive industries of China's masses whose life is toil and thus an account of Chinese civilization* (1937). Cambridge, Mass., and London, M.I.T. Press

Houel, J. P. 1782–7. *Voyage pittoresque aux isles de Sicile, de Malte, et de Lipari*. 4 vols., Paris, Imprimerie de Monsieur

Hoyos, F. A. 1978. *Barbados. A history from the Amerindians to independence*. London and Basingstoke, Macmillan Education

Hudson, Colin. 1984. The Barbados sugar industry fights back. *Sugar y Azúcar*, **79**, no. 12, 14–22

Huetz de Lemps, Alain. 1977. *La Canna à sucre au Brésil*. Travaux et Documents de Géographie Tropicale, no. 29, Talence, Domaine Universitaire de Bordeaux, Centre d'Etudes de Géographie Tropicale

Humboldt, Alexander von. 1811. *Political essay on the kingdom of New Spain*. Trans. John Black, 4 vols., London: Hurst, Rees, Orme and Brown; and H. Colborn; Edinburgh: W. Blackwood, Brown and Cambrie

1826. *Essai politique sur L'Ile de Cuba*. 2 vols., Paris, Librairie de Gide Fils

Ibn al-Awwam. 1802. *Libro de agricultura*. Trans. Don Josef Antonio Banqueri,

2 vols., Madrid, Imprensa Real
Imamuddin, S. M. 1963. Al-Filahah (farming) in muslim Spain. *Islamic Studies*, **1**, 51–89
Jackson, James C. 1968. *Planters and speculators. Chinese and European agricultural enterprise in Malaya, 1786–1921*. Kuala Lumpur and Singapore, University of Malaya Press
Johnson, H. 1972. The origins and early development of cane farming in Trinidad, 1882–1906. *Journal of Caribbean History*, **5**, 46–74
Jones, Philip. 1966. Medieval agrarian society. In M. M. Postan (ed.), *The Cambridge economic history of Europe*, vol. 1, pp. 340–431. Cambridge, Cambridge University Press
Keith, Robert K. 1976. *Conquest and agrarian change. The emergence of the hacienda system on the Peruvian coast*. Cambridge, Mass., and London, Harvard University Press
Kirchner, John A. 1980. *Sugar and seasonal labour migration. The case of Tucumán, Argentina*. Research Papers, no. 192, Chicago, Department of Geography, University of Chicago
Knight, F. W. 1977. Origins of wealth and the sugar revolution in Cuba, 1750–1850. *Hispanic American Historical Review*, **57**, 231–53
Knight, G. R. 1980. From plantation to padi-field: the origins of the nineteenth century transformation of Java's sugar industry. *Modern Asian Studies*, **14**, 177–204
Koster, Henry. 1816. *Travels in Brazil*. London, Longman, Hurst, Rees, Orme and Brown
Labat, J. P. 1742. *Nouveau voyage aux isles de L'Amérique, contenant l'histoire naturelle de ces pays, l'origine, les moeurs, la religion et la gouvernement de habitants anciens et modernes. Les guerres et les événements singuliers qui y sont arrivez pendant le séjour que l'auteur y a fait*. 8 vols., Paris, Chez Ch. J. B. Delespine
La Couture, Dutrône. 1790. *Précis sur la canne et sur les moyens d'en extraire le sel essentiel, suivi de plusieurs mémoires sur le sucre, sur le vin de sucre, sur l'indigo, sur les habitations et sur l'état actuel de Saint Domingue*. Paris, Duplain
Ladurie, Le Roy. 1971. *Times of feast, times of famine. A history of climate since the year 1000*. Garden City, N.Y., Doubleday and Co.
Laet, Jean de. 1640. *L'Histoire du nouveau monde, ou description des Indes Occidentales, contenant dix-huit livres, par le sieur Jean de Laet, d'Anvers; enrichi de nouvelles tables géographiques et figures des animaux, plantes et fruits*. Leyde, Chez Bonaventure et Abraham Elseviers
Lamb, H. H. 1965. The early medieval warm epoch and its sequel. *Paleogeography, Paleoclimatology, Paleoecology*, **1**, 13–37
Lapeyre, Henri. 1959. *Géographie de l'Espagne morisque*. Paris. S.E.V.P.E.N.
Larkin, John A. 1984. The international face of the Philippine sugar industry, 1836–1920. *Philippine Review of Economics and Business*, **21**, 39–58
Lasserre, Guy. 1961. *La Guadeloupe. Etude géographique*. 2 vols., Bordeaux, Union Française d'Impression

Laufer, Berthold. 1919. *Sino-Iranica. Chinese contributions to the history of civilization in ancient Iran; with special reference to the history of cultivated plants and products.* Anthropological Series, vol. 15, Chicago, Field Museum of Natural History

Laurence, K. O. 1971. *Immigration into the West Indies in the 19th century.* Kingston, Caribbean Universities Press, and Aylesbury, Ginn and Co.

Lemon, James T. 1972. *The best poor man's country. A geographical study of early southeastern Pennsylvania.* Baltimore, Johns Hopkins University Press

Léry, Jean de. 1586. *Historia navigationis in Brasiliam quae et America dicitur . . . Gallice scripta, nunc vero primum latinitate donata et variis figuris illustrata.* 8 vols., Genevae, apud Eustathium Vignon

Leslie, Charles. 1740. *A new and exact account of Jamaica, wherein the ancient and present state of that colony, its importance to Great Britain, laws, trade, manners and religion, together with the remarkable and curious animals, plants, trees etc. are described.* 3rd edn., Edinburgh, R. Fleming

Lévi-Provencal, E. 1967. *Histoire de l'Espagne musulmane.* 2 vols., Paris, Editions G. P. Maisonneuve et Larose

Levy, Claude. 1980. *Emancipation, sugar and federalism. Barbados and the West Indies, 1833–1876.* Gainesville, University of Florida Press

Lewis, W. A. (ed.). 1970. *Tropical development 1880–1913. Studies in economic progress.* London, Allen and Unwin

Leyburn, James G. 1966. *The Haitian people.* 2nd edn., New Haven, Yale University Press

Li, Hui-Lin. 1979. *Nan-fang ts'ao-mu chuang. A fourth century flora of southeast Asia.* Hong Kong, Chinese University Press

Ligon, Richard. 1657. *A true and exact history of the island of Barbados.* London, Humphrey Moseley

Lippmann, Edmund Oskar von. 1929. *Geschichte des Zuckers, seit den ältesten Zeiten bis zum Beginne der Rubenzucker-Fabrikation.* 2nd edn., Berlin, Julius Springer

Lockhart, James, and Schwartz, Stuart, B. 1983. *Early Latin America. A history of colonial Spanish America and Brazil.* Cambridge, Cambridge University Press

Lobdell, Richard A. 1972. Patterns of investment and sources of credit in the British West Indian sugar industry, 1838–97. *Journal of Caribbean History,* **4**, 31–53

Lombard, Maurice. 1957. Arsenaux et bois de marine dans la Méditerranée musulmane (VII–XI siècles). In *2nd colloque internationale d'histoire maritime, 1957: le navire et l'économie maritime du moyen âge au XVIII siècle principalement en Méditerranée,* pp. 53–99. Paris, S.E.V.P.E.N.

 1959. Un problème cartographié: le bois dans la Méditeranée musulmane (VII–XI siècles). *Annales: Economies, Sociétés, Civilisations,* **14**, 234–54

Lopez, Fernando Charadan. 1982. *La industria azucarera en Cuba.* La Habana, Editorial de Ciencias Sociales

Lopez, Robert. 1970. Italy (part of Robert Lopez, Harry Miskimin and Abraham Udovitch, "England to Egypt, 1300–1500: long-term trends and long-distance trade"). In M. A. Cook (ed.), *Studies in the economic history of the Middle*

East, pp. 107–15. London, Oxford University Press
Lowndes, A. G. 1956. The sugar industry of Fiji. In A. G. Lowndes (ed.), *South Pacific enterprise. The Colonial Sugar Refining Company Limited*, pp. 67–90. Sydney, Angus and Robertson
Lyle, Philip. 1950. The sources and nature of statistical information in special fields of statistics. The sugar industry. *Journal of the Royal Statistical Society.* (Series A), **113**, 531–43
McCusker, John J. 1973. Weights and measures in the colonial sugar trade: the gallon and the pound and their international equivalents. *William and Mary Quarterly*, 3rd ser., **30**, 599–624
Macera, Pablo. 1974. *Las plantaciónes azucareras en el Perú, 1821–1875.* Lima, Biblioteca Andina
McManis, Douglas. 1975. *Colonial New England. A historical geography.* New York, Oxford University Press
McPhee, John. 1967. *Oranges.* New York, Farrar, Strauss and Giroux
Maity, Sachindra Kumar. 1970. *Economic life in northern India in the Gupta period (c. A.D. 300–550).* 2nd edn., Delhi, Varanasi, Patna, Motilal Banarsidass
Malowist, Marian. 1969. Les Débuts du système des plantations dans la période des grandes découvertes. *Africana Bulletin*, **1**, 9–30
Mannick, A. R. 1979. *Mauritius. The development of a plural society.* Nottingham, Spokesman
Marcgravius de Liebstad, Georgius. 1648. *Historiae rerum naturalium Brasiliae*, part of Gulielmus Piso, *Historia naturalis Brasiliae.* Lugdun. Batavorum, apud Franciscum Hackium et Amstelodami, apud Lud. Elzevirium
Marques, A. H. de Oliveira. 1972. *History of Portugal.* 2 vols., New York and London, Columbia University Press
Marrero y Artiles, Levi. 1972–84. *Cuba: economía y sociedad.* 11 vols., Madrid, Editorial Playor
Marshall, Woodville, K. 1975. The establishment of a peasantry in Barbados, 1840–1920. In T. Matthews (ed.), *Social groups and institutions in the history of the Caribbean*, pp. 84–104. Puerto Rico, Association of Caribbean Historians
Martin, Samuel. 1765. *An essay on plantership, humbly inscribed to his excellency George Thomas, esq., Chief Governor of all the Leeward Islands.* 4th edn., Antigua, Samuel Clapham, and London, A. Miller
Masefield, G. B. 1967. Crops and livestock. In E. E. Rich and C. H. Wilson (eds.), *The Cambridge economic history of Europe*, vol. 4, pp. 276–301. 7 vols., Cambridge, Cambridge University Press
Mattei, Andres A. Ramos. 1984. The growth of the Puerto Rican sugar Industry under North American domination: 1899–1910. In Albert and Graves, pp. 121–31
Matznetter, Josef. 1958. *Die Kanarischen Inseln, Wirtschaftsgeschichte und Agrargeographie.* Petermanns Geographischen Mitteilungen, no. 266, Gotha, Hermann Haack
Maude, H. E. 1981. *Slavers in paradise. The Peruvian labour trade in Polynesia, 1862–1864.* Canberra, Australian National University Press
Mauro, Frédéric. 1960. *Le Portugal et l'Atlantique au XVIIe siècle.* Paris, S.E.V.-

P.E.N.
Meade, George P., and Chen, James C. P. 1977. *Cane sugar handbook. A manual for cane sugar manufacturers and their chemists*. 10th edn., New York, John Wiley
Melville, Roberto. 1979. *Crecimiento y rebelión. El desarollo económico de las haciendas azucareras en Morelos (1880–1910)*. México, Editorial Nueva Imagen
Merrill, Gordon C. 1958. *The historical geography of St. Kitts and Nevis, the West Indies*. México, Instituto Panamericano de Geografia e Historia
    1964. The role of the Sephardic Jews in the British Caribbean area during the seventeenth century. *Caribbean Studies*, **4**, 32–40
Mintz, Sidney, W. 1985. *Sweetness and power. The place of sugar in modern history*. New York, Elizabeth Sifton Books, Viking
Moore, Brian L. 1975. The social impact of Portuguese immigration into British Guiana after emancipation. *Boletin de Estudios Latinoamericanos y del Caribe*, **19**, 3–15
Morales Padron, Francisco. 1952. *Jamaica española*. Sevilla, Escuela de Estudios Hispano-Americanas
Moreton, J. B. 1793. *West India customs and manners: containing strictures on the soil, cultivation, produce, trade, officers, and inhabitants; with the methods of establishing and conducting a sugar plantation. To which is added the practice of training new slaves*. London, J. Parsons, W. Richardson, H. Gardner and J. Walter
Morimoto, Amelia. 1979. *Los inmigrantes japoneses en el Perú*. Lima, Universidad Nacional Agraria
Morison, Samuel Eliot. 1942. *Admiral of the ocean sea. A life of Christopher Columbus*. 2 vols., Boston, Little, Brown
Moya Pons, Frank. 1972. *La Dominación Haitiana 1822–1844*. 2nd edn., Santiago, Dominican Republic, Universidad Católica Madre y Maestra
Moynagh, Michael. 1981. *Brown or white? A history of the Fiji sugar industry 1873–1973*. National University of Australia Research Monographs, vol. 5, Canberra, Australian National University
Muffett, Dr. Thomas. 1655. *Health's improvement*. London, T. Newcomb for S. Thomson
Naqui, Hamida Khatoon. 1986. *Agricultural, industrial and urban dynamism under the sultans of Delhi, 1206–1555*. New Delhi, Munshiram Manoharlal Publishers Pvt. Ltd.
Needham, Joseph. 1965. *Physics and physical technology*, vol. 4, part 2, in Joseph Needham, *Science and civilization in China*. Cambridge, Cambridge University Press
Nef, John U. 1950. *War and human progress. An essay on the rise of industrial civilization*. London, Routledge and Kegan Paul
Nichols, Thomas. 1963. *A pleasant description of the fortunate islands called islands of Canaria with their strange fruits and commodities* (London, Thomas East, 1583). Reprinted in Alejandro Cioranescu, *Thomas Nichols, mercador de azúcar, hispanista e hereje*. La Laguna de Tenerife, Instituto de estudios Canarios
Ormrod, Richard K. 1979. The revolution of soil management practices in early

Jamaican sugar planting. *Journal of Historical Geography*, **5**, 157–70
Osborn, R. F. 1964. *Valiant harvest, the founding of the South African sugar industry, 1848–1926.* Durban, South African Sugar Association
Ott, Thomas O. 1973. *The Haitian revolution 1789–1804.* Knoxville, University of Tennessee Press
Oviedo y Valdés, Gonzalo Fernández de. 1959. *Historia general y natural de las Indias.* Edited and with an introduction by J. Pérez de Tudela Bueso, 5 vols., Madrid, Real Academia Espanola
Panebianco, Gaspar Vaccaro e. 1825–6. *Sul richiano della canna zuccheriora in Sicilia.* 2 vols., Palermo and Girgenti, Tipografia di Pipomi
Pares, Richard. 1950. *A West India fortune.* London, Longmans, Green and Co.
  1960. *Merchants and planters.* Economic History Review Supplement, no. 4, Cambridge, Cambridge University Press
Parreira, Henrique Gomes de Amorim. 1952. *História do açúcar em Portugal, Anais: estudos da história da geografia da expansão Portuguesa.* Lisbon, Junta das Missões Geograficas e de Investigações do Ultramar
Parry, John H. 1955. Plantation and provision ground. An historical sketch of the introduction of food crops into Jamaica. *Revista da Historia da America*, **39**, 1–20
Parry, J. H., and Sherlock, P. M. 1956. *A short history of the West Indies.* London, Macmillan
Parsons, James J. 1983. The migration of Canary islanders to the Americas: an unbroken current since Columbus. *The Americas*, **39**, 447–81
Pegolotti, Balducci. 1936. *La practica della mercatura*, ed. Allan Evans. Cambridge, Mass., Medieval Academy of America
Pereira, Soares Moacyr. 1955. *A origem dos cilindros na moagem da cana: investigação em Palermo.* Rio de Janeiro, Instituto do Açúcar e do Alcool
Perkins, John. 1984. The political economy of sugar beet in imperial Germany. In Albert and Graves, pp. 31–45
Petrone, Maria Thereza Schorer. 1968. *A lavoura canavieira em São Paulo. Expansão e declínio (1765–1851).* São Paulo, Difusão Européia do Livro
Pike, Ruth. 1966. *Enterprise and adventure. The Genoese in Seville and the opening of the New World.* Ithaca, Cornell University Press
Pinho, Wanderley. 1946. *História de um engenho do Recôncavo 1552–1944.* Rio de Janeiro, Livraria Editôra Zélio Valverde S.A.
Piso, Gulielmus. 1648. *Historia naturalis Brasiliae.* Lugdun. Batavorum, apud Franciscum Hackium et Amstelodami, apud Lud. Elzervirium
Pitcher, P. W. 1909. *In and about Amoy.* Shanghai and Foochow, Methodist Publishing House in China
Pluchon, Pierre (ed.). 1982. *Histoire des Antilles et de la Guyane.* Toulouse, Privat
Poliak, A. N. 1936. La Feodalité islamique. *Revue des Etudes Islamiques*, **10**, 247–65
Pomet, Pierre. 1694. *Histoire générale des drogues.* Paris, J.-B. Loyson
Pontes, Felisberto Caldeira Brant, Marquês de Barbacena. 1976. *Economia açucareiro do Brasil no século XIX.* Rio de Janeiro, Instituto do açúcar e do Álcool
Potter, Jim. 1984. Demographic development and family structure. In Jack P. Greene and J. R. Poole (eds.), *Colonial British America. Essays in the new*

history of the early modern era, pp. 123–56. Baltimore, Johns Hopkins University Press

Prawer, Joshua. 1952, 1953. Etude de quelques problèmes agraires et sociaux d'une seigneurie croisée au X111e siècle. *Byzantion*, **22**, 5–61; **23**, 143–70

1972. *The Latin Kingdom of Jerusalem. European colonization in the middle ages.* London, Weidenfeld and Nicolson

Price, Richard (ed.). 1979. *Maroon societies: rebel slave communities in the Americas.* 2nd edn., Baltimore, Johns Hopkins University Press

Quigley, N. C. 1985. *The Bank of Nova Scotia in the Caribbean 1889–1940. Canadian banking within the American commercial empire.* Wellington, New Zealand, Department of Economic History, Victoria University

Rabie, Hassanein. 1972. *The financial system of Egypt A.H. 564–741/A.D. 1169–1341.* London, New York and Toronto, Oxford University Press

Ratekin, Mervyn. 1954. The early sugar industry in Española. *Hispanic American Historical Review*, **34**, 1–19

Rau, Virgínia. 1964. The settlement of Madeira and the sugar cane plantations. *Afdeling Agrarische Geschiedenis Bijdragen*, **11**, 3–12

Rau, Virgínia, and Macedo, Jorge de. 1962. *O açúcar da Madeira nos fins do século XV: problemas de produção e comércio.* Funchal, Junta-Geral do Distrito Autonomo do Funchal

Rawski, Evelyn S. 1972. *Agricultural change and the peasant economy of South China.* Cambridge, Mass., Harvard University Press

Raychaudhuri, S. P. 1964. *Agriculture in ancient India.* New Delhi, Indian Council of Agricultural Research

Raychaudhuri, Tapan. 1982. Mughal India. In Tapan Raychaudhuri and Irfan Habib (eds.), *The Cambridge economic history of India, vol. 1, c. 1200–c. 1750*, pp. 261–307. Cambridge, Cambridge University Press

1983. The mid-eighteenth century background. In Dharma Kumar (ed.), with the editorial assistance of Meghdad Desai, *The Cambridge economic history of India, vol. 2, c. 1757–c. 1970*, pp. 3–35. Cambridge, Cambridge University Press

Renaud, H. P. J. 1948. *Le Calendrier d'Ibn al-Banna de Marrakech (1256–1321 A.D.).* Paris, Larose Editeurs

Revert, Eugène. 1949. *La Martinique. Etude géographique et humaine.* Paris, Nouvelles Editions Latines

Richardson, Bonham C. 1983. *Caribbean migrants. Environment and human survival on St. Kitts and Nevis.* Knoxville, University of Tennessee Press

1985. *Panama money in Barbados, 1900–1920.* Knoxville, University of Tennessee Press

Richardson, Peter. 1984. The Natal sugar industry in the nineteenth century. In Albert and Graves, pp. 237–58

Riley-Smith, Jonathan. 1967. *The knights of St. John in Jerusalem and Cyprus, c. 1050–1310.* London, Macmillan

1973. *The feudal nobility and the kingdom of Jerusalem, 1174–1277.* London, Macmillan

Robinson, D. J., and the Editors. 1983. The Guianas. In Harold Blakemore and

Clifford Smith (eds.), *Latin America. Geographical perspectives*, pp. 241–52. 2nd edn., London, Methuen

Rouse, I. 1949. The West Indies. In Julian H. Steward (ed.), *The handbook of South American indians*, vol. 4, pp. 495–565. Washington, Smithsonian Institution

Ruiz, Juan Martínez. 1964. Notas sobre el refinado del azúcar de caña entre los moriscos granadinos. Estudio lexico. *Revista de dialectologia y tradiciones populares*, **20**, 271–88

Saint-Méry, Médéric Louis Elie Moreau de. 1797–8. *Description topographique, physique, civile, politique et historique de la partie française de l'isle de Saint-Domingue*. 2 vols., Philadelphia, n.p.

Salvador, Frei Vicente do. 1965. *História do Brasil 1500–1627*. 5th edn., São Paulo, Edições Melhoramentos

Sánchez-Albornoz, Nicolás. 1984. The population of colonial Spanish America. In Leslie Bethell (ed.), *The Cambridge history of Latin America*, vol. 2, pp. 3–35. Cambridge, Cambridge University Press

Sandoval, Fernando B. 1951. *La industria del azúcar en Nueva España*. Publicaciones del Instituto de Historia, no. 21, México, Universidad Nacional Autonoma de México

Sauer, Carl Orwin. 1966. *The early Spanish Main*. Berkeley and Los Angeles, University of California Press

Saunders, Kay. 1982. *Workers in bondage. The origins and bases of unfree labour in Queensland 1824–1916*. St. Lucia, University of Queensland Press

Scarano, Francisco A. 1984. *Sugar and slavery in Puerto Rico. The plantation economy of Ponce, 1800–1850*. Madison, University of Wisconsin Press

Schafer, E. H. 1963. *The golden peaches of Samarkand. A study of T'ang exotics*. Berkeley and Los Angeles, University of California Press

  1977. T'ang. In K. C. Chang (ed.), *Food in Chinese culture. Anthropological and historical perspectives*, pp. 87–140. New Haven and London, Yale University Press

Scherer, André. 1980. *La Réunion*. Paris, Presses Universitaires de France

Schmitz, Mark. 1979. The transformation of the southern cane sugar sector, 1860–1930. *Agricultural History*, **53**, 270–85

Schnakenbourg, Christian. 1980. *Histoire de l'industrie sucrière en Guadeloupe au XIXième et XXième siècles: la crise du système escalavagiste, 1835–1847*. Paris, L'Harmattan

  1984. From sugar estate to central factory: the industrial revolution in the Caribbean (1840–1905). In Albert and Graves, pp. 83–94

Schwartz, Stuart B. 1978. Indian labor and new world plantations: European demands and Indian responses in northeast Brazil. *American Historical Review*, **83**, 43–79

Scott, Rebecca, J. 1984. The transformation of sugar production in Cuba after emancipation, 1880–1900: planters, *colonos* and former slaves. In Albert and Graves, pp. 111–19

Serrão, Joel. 1954. Le Blé des isles Atlantiques: Madère et Açores aux XVe et XVIe siècles. *Annales: Economies, Sociétés, Civilisation*, **9**, 337–41

Shafer, R. 1958. *Economic societies in the Spanish world, 1763–1821*. Syracuse, Syracuse University Press
Sheppard, Jill. 1977. *The "Redlegs" of Barbados*. Millwood, New York, K.T.O. Press
Sheridan, Richard B. 1960. Samuel Martin, innovating sugar planter of Antigua 1750–1776. *Agricultural History*, **34**, 126–39
  1961. The rise of a colonial gentry. A case study of Antigua, 1730–1775. *Economic History Review*, 2nd ser., **13**, 342–57
  1965. The wealth of Jamaica in the eighteenth century. *Economic History Review*, 2nd ser., **18**, 292–311
  1974. *Sugar and slavery. An economic history of the British West Indies*. Baltimore, Johns Hopkins University Press
Silva, Helder Lains e. 1958. *São Tomé e Principé e a cultura do cafe*. Memórias da Junta de Investigações do Ultramar, no. 1, 2nd ser., Lisbon, Ministério do Ultramar
Simmons, Adele Smith. 1982. *Modern Mauritius. The politics of decolonization*. Bloomington, Indiana University Press
Sitterson, J. Carlyle. 1953. *Sugar country: the cane sugar industry in the South, 1753–1950*. Lexington, University of Kentucky Press
Slicher van Bath, B. H. 1963. *The agrarian history of western Europe A.D. 500–1850*. London, Edward Arnold
Sloane, Sir Hans. 1707. *A voyage to the islands Madeira, Barbados, Nieves, S. Christophers and Jamaica, with the natural history of the herbs and trees, four-footed beasts, fishes, birds, insects, reptiles, etc. on the last of those islands*. London, printed by B. M. for the Author
Soares de Sousa, Gabriel. 1971. *Tratado descritivo do Brasil em 1587*. 4th edn., São Paulo, Companhia Editôra Nacional and Editôra da Universidade de São Paulo
Spalding, Thomas. 1816. *Observations on the method of planting and cultivating sugar cane in Georgia and South Carolina*. Charleston, Agricultural Society of South Carolina (Reprinted in E. Merton Coulter (ed.), *Georgia's disputed ruins*, Chapel Hill, University of North Carolina Press, 1937, pp. 237–63)
Stevenson, G. C. 1965. *Genetics and breeding of sugar cane*. London, Longmans
Steward, Julian H., and Faron, Louis C. 1959. *Native peoples of south America*. New York, McGraw-Hill
Stewart, Watt. 1951. *Chinese bondage in Peru. A history of the Chinese coolie in Peru, 1849–1874*. Durham, North Carolina, Duke University Press
Straet, Jan van der. 1600. *Nova Reperta*. Antuerpia, Philippus Gallaeus
Sung Ying-Hsing. 1966. *T'ien-Kung K'ai-Wu. Chinese technology in the seventeenth century*. Trans. E-Tu Zen Sun and Shiou-Chuan Sun, University Park and London, Pennsylvania State University Press
Tate, Merze. 1968. *Hawaii. Reciprocity or annexation*. East Lansing, Michigan State University Press
Tenreiro, Francisco. 1961. A Ilha de São Tomé. *Memórias da Junta de Investigações do Ultramar*, **24**, 67–74
Tertre, Jean Baptiste du. 1654. *Histoire générale des isles de S. Christophe, de*

la Guadeloupe, de la Martinique, et autres dans l'Amérique. Paris, chez Jacques Langlois, Imprimeur Ordinaire du Roy, et Emmanuel Langlois

Thiriet, Freddy. 1959. *La Romanie vénitienne au moyen âge: le développement et l'exploitation du domaine colonial vénitien (XIIe–XVe siècles)*. Paris, Bibliothèque des Ecoles Françaises d'Athènes et du Rome, Editions E. de Boccard

1967. La Condition paysanne et les problèmes de l'exploitation rurale en Romanie greco-vénitienne. *Studi Veneziani*, 9, 35–71

Thomas, Sir Dalby. 1690. *An historical account of the rise and growth of the West India colonies*. London, printed for J. Hindmarsh

Thomas, Hugh. 1971. *Cuba or the pursuit of freedom*. London, Eyre and Spottiswoode

Thomas, Mary Elizabeth. 1974. *Jamaica and voluntary laborers from Africa, 1840–1865*. Gainesville, University Presses of Florida

Timoshenko, V. P., and Swerling, B. C. 1957. *The world's sugar. Progress and policy*. Stanford, Stanford University Press

Tinker, Hugh. 1974. *A new system of slavery. The export of Indian labour overseas 1830–1920*. London, Oxford University Press

Toussaint, Auguste. 1972. *Histoire des Iles Mascareignes*. Paris, Berger-Levrault

Trasselli, Carmelo. 1957. Producción comercio de azúcar en Sicilia del siglo XIII al XIX. *Revista Bimestre Cubana*, 72–3, 130–54

Turner, M. 1974. Chinese contract labour in Cuba 1847–74. *Caribbean Studies*, 14, 66–81

Udovitch, Abraham. 1970. Egypt (part of Robert Lopez, Harry Miskimin and Abraham Udovitch, "England to Egypt, 1300–1500: long-term trends and long-distance trade." In M. A. Cook (ed.), *Studies in the economic history of the Middle East*, pp. 115–28. London, Oxford University Press

Vanacker, Claudette. 1973. Géographie économique de l'Afrique du Nord selon les auteurs arabes, du IXe siècle au milieu du XIIe siècle. *Annales: Economies, Sociétés, Civilizations*, 28, 659–80

Vellard, J. 1939. *Une civilization du miel. Les indiens Guayakis du Paraguay*. Paris, Librairie Gallimard

Verlinden, Charles. 1970. *The beginnings of modern colonization*. Ithaca and London, Cornell University Press

Wacker, Peter O. 1975. *Land and people. A cultural geography of pre-industrial New Jersey: origins and settlement patterns*. New Brunswick, N.J., Rutgers University Press

Wagner, Michael J. 1977. Rum, policy and the Portuguese: or the maintenance of elite supremacy in post-emancipation British Guiana. *Canadian Revue of Sociology and Anthropology*, 14, 406–16

Wallerstein, Immanuel. 1980. *Mercantilism and the consolidation of the European world-economy, 1600–1750*, vol. 2 in *The modern world system*. New York, Academic Press

Warman, Arturo. 1984. The cauldron of the revolution: agrarian capitalism and the sugar industry in Morelos, Mexico, 1880–1910. In Albert and Graves, pp. 165–79

Wartburg, Marie-Louise von. 1983. The medieval cane sugar industry in Cyprus:

results of recent excavations. *Antiquaries Journal*, **63**, 298–314
Watson, Andrew M. 1974. The Arab agricultural revolution and its diffusion, 700–1100. *Journal of Economic History*, **34**, 8–35
  1983. *Agricultural innovation in the early Islamic world*. Cambridge, Cambridge University Press
Watts, David. 1968. Origins of Barbadian cane hole agriculture. *Barbados Museum and Historical Society Journal*, **32**, 143–51
White, Lynn. 1962. *Medieval technology and social change*. Oxford, Clarendon Press
Whiteford, Scott. 1981. *Workers from the north. Plantations, Bolivian labor and the city in northwest Argentina*. Austin, University of Texas Press
Wigginton, Eliot (ed.). 1975. *Foxfire 3*. Garden City, New York, Anchor Press/Doubleday
Williams, D. H. 1974. Cistercian settlement in the Lebanon. *Citeaux. Commentari Cistercienses*, **25**, 61–74
Williams, Eric. 1964. *History of the people of Trinidad and Tobago*. André Deutsch, London
Williams, Jack F. 1980. Sugar: the sweetener in Taiwan's development. In Ronald G. Knapp (ed.), *China's island frontier. Studies in the historical geography of Taiwan*, pp. 219–51. Honolulu, University of Hawaii Press and the Research Corporation of the University of Hawaii
Willis, Geoffrey A. 1972. *Harvesting and transporting sugar cane in Australia*. James Cook University Monograph Series, no. 3, Townsville, James Cook University of North Queensland
Wilson, C. Anne. 1985. *The Book of Marmalade*. London, Constable
Winberry, John J. 1980. The sorghum syrup industry 1854–1975. *Agricultural History*, **54**, 343–52
Wiznitzer, Arnold. 1960. *Jews in colonial Brazil*. New York, Columbia University Press
Womack, John Jr. 1969. *Zapata and the Mexican revolution*. New York, Alfred A. Knopf
Wood, Donald. 1968. *Trinidad in transition. The years after slavery*. London and New York, Oxford University Press
Yü, Ying-shih. 1977. Han. In K. C. Chang (ed.), *Food in Chinese culture. Anthropological and historical perspectives*, pp. 53–83. New Haven and London, Yale University Press

Index

Abu Hanifa, 33
Abyssinia, first mention of sugar cane in, 24
Africa, diffusion of sugar cane to, 24, 33–4
agent(s): functions of, 111; as source of capital, 146, 178; as owners, 179; in Hawaii, 226–7
agricultural manuals, 27, 56, 94, 204, 232
Al-Nuwairi, 38, 39
America, *see* United States of America
American Sugar Refining Company (Sugar Trust), 167
Americas
  sixteenth and seventeenth centuries: introduction of sugar cane to, 62–3; adaption of industry to new conditions, 63–4; early industry in, 64–83; *see also* Barbados; Brazil; Hispaniola
  eighteenth century: general conditions, 84–8; cost of plantations, 88–90; land use, 90–3; innovations, 93–105; mills, 105; manufacture of sugar, 105–10; sugar trade, 110–12; plantation society, 112–19; population, 113–17; towns, 118; *see also* sugar (manufacture of); sugar mills and presses; and *under* individual colonies
  nineteenth century: cane breeding, 123, 141–2; competition from beet, 120, 130, and from new cane sugar producers, 120; labor problem, 123–30; abolition of slavery, 121–2, 123–4; new technology, 122–3, 134–41; *see also under* individual countries and colonies
Annam, early industry in, 21
Annamese labor, 129
Antigua, 124, 129, 133, 225; cane-holes on, 102; windmills on, 102; abolition, and reaction to, 154; technical improvements and central factories, 154–5; decline, 236

Antonil, André João, 107, 108
Arab agricultural revolution, 27, 33, 119
Arabia, diffusion of sugar cane through, 24
Arabs, role in diffusion of sugar cane, 23, 25–7, 33–4
Arango y Pareño, Francisco, 163, 164
Argentina, 144; nationalist policies, 186, 187–8; Argentinian ownership, 187; French interest, 187; labor, 187; production, 186; railways, 186, 187; refinery built, 188
arrobas (Spanish), xiii
arrôbas (Portuguese), xiii
Ashtor, Eliyahu, 55, 111
aspartame, 5
Australia: distinctiveness of, 228–9; labor policies, 129, 229–30, 231–2; cane-farmers, 231; unions, 232; central factories, 230–1; tariffs for, 232; Melanesian canes in, 141; research, 141, 232–3
Azores, unsuitable for sugar cultivation, 50

bagasse: around Mediterranean, 39; with the three-roller mill, 75, 97; as fuel, 90, 97–9, 206, 235; as raw material for industry, 235
Bahia, *see* Brazil
Barbados: introduction of monoculture, 80–1; social revolution in, 80–2; early prosperity of, 82, 86; emigration from, 81, 130, 151–2; slavery in, 81, 115; bagasse for fuel, 98; use of manure, 99; erosion in, 100; cane-holes on, 100–2; windmills on, 102; as center of innovation, 100; population of, 113; type of sugar made, 110, 153; abolition, and reaction to, 144, 150; Court of Chancery, 152; Encumbered Estates

259

260 Index

Barbados (*cont.*)
    Act, 152; cost of plantations on, 88, 163, and sale of, 152, and amalgamation of, 153; wages in, 150, 153; remittances (and "Panama money"), 152; smallholders, 152; new villages, 152; cane breeding, 142, 150; steam mills, 135; slow adoption of new technology, 150, 153–4; central factories, 154; ownership of industry, 158; future, 236
beet, *see* sugar beet; sugar beet industry
Bengal, *see* India
Booker Bros., McConnell and Co., 179
botanical gardens, 25, 27, 96, 141
Bourbon cane, *see* sugar cane (cultivated varieties)
Braudel, Fernand, 6, 31
Brazil
    northeastern (including Bahia, Pernambuco): plantations (*engenhos*), 71–3, 77; *senhor de engenho*, 71, 73, 75; *lavrador*, 71–2; Indians, 72; slaves, 72, 115, 116; land use, 72–3; mills, 73; three-roller mill, 39, 73–6; innovation in manufacturing, 76–7; Dutch in, 54, 78; diffusion of industry to Barbados, 77–80; bagasse for fuel, 99; manufacture of sugar, 107; population, 113, 115, 116; frontier, 112, 160; Palmares, 117; expansion in nineteenth century, 158–60; abolition, and reaction to, 160; *moradores*, 160, 161; steam mills, 135; central factory system, 161; comparison with Peru, 182; *rapadura*, 139, 182; exports to Mediterranean, 45; internal markets, 162; for alcohol fuel, 235
    Rio de Janeiro, 84, 85, 112, 113, 193
    São Paulo, 130, 144, 182, 193
breadfruit, 94, 104
British Guiana, 86, 130, 144, 163, 173; drainage, 103; maroons, 117; policies towards ex-slaves, 174–5; Madeirans in, 125, 175–6; migrant labor, 175; indentured Indian labor, 126–7, 176–7; steam power, 179; vacuum pan installed, 137, 179; sources of capital, 178–9; change in ownership of plantations, 179–80; Booker Bros., McConnell and Co., 179
Brussels Convention of 1902, 10, 134, 148, 180, 188, 194
Bry, Theodor de, 67, 68, 75

Canadian investment: in Cuba, 168; in Fiji, 226
Canarians: in Hispaniola, 65, 66; Dominican Republic, 173

Canary Islands: discovery, 55; introduction of industry, 32, 55–6; influence of Madeira, 55, 57; cultivation, 56–7; labor, 56–7; mills, 57–8; decline of industry, 58; sources for, 80–1
cane-farmers, 72, 81, 177–8, 209, 224, 229, 230, 231, 235
cane-holes, 100–2
cane sugar, *see* sugar cane
Cape Verde Islands, unsuitable for sugar industry, 48
Caribbean, *see* Americas; and *under* individual islands
Caspian Sea, 24
Castro, Fidel, 16
Cauca Valley, *see* Colombia
centers of innovation, 18; northern India, 19–21; Tonkin, 22; southern Spain, 33; São Tomé, 58, 59, 60; Brazil, 72–7, 93; Barbados, 93, 100, 101; Java 213; Hawaii, 228
central factories, and central factory system, *see* sugar (manuacture of); and *under* individual countries and colonies
centrifugals, *see* sugar (manufacture of)
Charles Hires Company, 168
chemistry, growing importance for industry, 138
China, 198, 208, 209; origins of industry, 21–2; in cuisine, 22–3, 203–4; cultivation in Fukien, Kwangtung, Szechwan, 204–5; two-roller mill, 205–6, 207; three-roller mill, 207; types of sugar, 206–7; refineries, 207; reasons for lack of change, 207–8; Hong Kong merchants, 207
Chinese labor, *see* labor
clayed sugar, *see* sugar (types of)
Colombia, 144, 182, 193
Colonial Company, 179
Colonial Sugar Refining Company: in Australia, 230–1; in Fiji, 225–6; role in research, 228, 232
Columbus, Christopher, 54, 61, 63
Creole cane, *see* sugar cane (cultivated varieties)
Crete: beginnings of industry, 34; slavery, 42, 119; increase in production, 42, 44–5; decline of industry, 45; *see also* Mediterranean sugar industry
Crusader states, 27, 39; landownership in, 41; decline of industry, 43; *see also* Mediterranean sugar industry
Cuba: earliest years, 70; slow to adopt innovations, 98, 99, 100; frontier, 112; population, 113; maroons, 117; slavery, 115, 165; steam mills, 119, 135; benefits

from revolt in St. Domingue, 163, 164;
Otaheiti cane in, 164; Golden Age to
1868, 164–5; consequences of the Ten
Years' War, 164, 165, 166, 167, 172;
railways, 164; abolition, 165; Chinese
labor, 128, 164, 165; *colonato* and
colonos, 165–6; migrant labor (from
Spain, Haiti, Jamaica), 129, 166–7;
patronato, 165; central factory system,
166, 167; American investment in,
167–8
Cuba Dominican Sugar Company, 173
Cuban Cane Sugar Corporation, 168
Cyprus, 27, 31; beginnings of industry, 34;
slavery, 42, 119; Cornaro family, 42;
increase in production, 42, 44–5; decline
of industry, 45; *see also* Mediterranean
sugar industry
Czarnikow Ltd., 168

Damascus, 24
David, Elizabeth, use of sugar in baking, 71
Decauville, 187, 228
Deer, Noel, 23, 33, 73, 74; use of statistics,
xiii
defrutum, 2
Demerara, *see* British Guiana
Digby, Sir Kenelm, 6
Dominica, 112
Dominican Republic, 144, 162; early
nineteenth century, 170–2; revival of
industry, 172–3; foreign investment in,
171, 172–3; price of land in, 163, 172;
railways, 172; *colono* system and Cuban
influence, 172; migrant labor, 130, 173;
see also Hispaniola
Dutch, 54, 61, 76, 77, 78, 79, 80, 83, 110,
197, 204, 209; *see also* Java

East India Company (British), 6, 199
East India Company (Dutch), 210
Edwards, Bryan, 63, 89, 100, 103
Egypt, 24, 33; first reference to sugar cane,
24; Mamluk landownership, 41–2;
decline of industry, 43–4; *see also*
Mediterranean sugar industry
Encumbered Estates Act (West Indies), *see
under* individual British West Indian
colonies
engenho, *see* Brazil; sugar mills and presses
Essay on Plantership, see Martin, Samuel
Euphrates, valley of, 24

factor, *see* agent(s)
Farmer, Fannie, use of sugar in baking,
salads, 8, 9

Fiji: markets in Australia, New Zealand,
Canada, 219; decline in Fijian
population, 222; indentured Indians,
222–3, 225; central factory system,
225–6
free trade, consequences for British sugar
colonies of, 120, 133–4, 144
French West India Company, 112
Freyre, Gilberto, 140
fuel: shortage of, 17, 39, 93, 97–9; *see also*
bagasse

Gage, Thomas, 62
Genoa (Genoese): role in Spain (Granada),
45; Portugal, 34: Madeira, 45, 54;
Canaries, 45; São Tomé, 60; Hispaniola,
65; Americas, 110
Gladstone, John, 126, 137, 179
Glasse, Hannah, Mrs., 7
Gopal, Lallanji, 20
Grace and Company, 182
Grainger, James, *The sugar-cane*, 99
grape juice, *see* must
Greater Antilles, *see under* individual islands
Grenada, 86, 112
Guadeloupe, 62, 79, 82, 86, 98, 114;
windmills on, 102; competition with
beet, 156; post-abolition labor, 156–7;
early introduction of central factory
system, 157–8; future, 236
Guinea grass, introduction of, 104

Hadramawt, early presence of sugar cane in,
24
Haiti: independence of, 144, 145; destruction
of industry, 145–6; *see also* St.
Domingue
Havana, 86
Hawaii, *see* United States of America
Hernandez, Francisco, agricultural manual
of, 56
Hershey's chocolate, 168
Heyd, Wilhelm von, 31
high fructose corn syrup (HFCS), 5, 235, 236
Hispaniola: beginnings of industry, 64–6;
European foremen, 66; mills and
manufacture, 65–6, 67–8; decline, 68–9;
population, 113; *see also* Dominican
Republic; Haiti; St. Domingue
honey, 4, 5, 22, 203
Humboldt, Alexander von, 96

Ibn al-Awwam, 36
Ibn Hawqal, 33
indentured labor, *see* labor (indentured)
India, 23, 25; state of industry and regions
of production *c*. 1750, 198–9, 201–2;

Index

India (*cont.*)
    imports sugar, 201; nature of domestic market, 201; technologically conservative, 202–3; central factory system, 202
    Bengal, 198, 199; East India Company encourages, 199, 201
    northern, 18, 23; early industry from pali and Sanskrit sources, 19–21; diffusion of industry eastwards, 21–2, and to Africa, 24, and westwards to Mediterranean, 23–7
    see also centers of innovation
Indo-China: early industry in, 21–2; *see also* Annam; centers of innovation; Tonkin
ingenio, see sugar mills and presses
innovations, *see* centres of innovation
Iran, 22, 23, 27
irrigation: significance to early Middle Eastern industry, 18, 25, 27; around the Mediterranean, 33, 36; on Madeira, 50; on Canary Islands, 55, 58; lack of in early American industry, 63; on St. Domingue, 102–3; in Mexico, 103; in Peru, 103; in Malaya, 216; on Hawaii, 228
Islam, *see* Arabs
Italian investment, 172, 173

Jamaica: under Spain, 70; English annexation, 82; cost of plantations, 89; fuel, 98–9, 147; manure, 99, 147; population, 113; slavery, 115; maroons, 117; consequences of and response to abolition, 146–7; shortage of capital, 147; Encumbered Estates Court, 147–8; decline in production, 148; central factory system, 148; emergence of a peasantry, 148–9
Jamaica train: advantages of, 97–8; diffusion of, 99
Japan: increase in consumption, 197, 208; destruction of indigenous industry following Meiji Restoration, 208; colonisation of Taiwan, 209; *see also* Taiwan
Japanese laborers, 129, 181, 223, 230
Java: arrival of Dutch, 210; Chinese technology in, 210–11; culture system, 211–12; post-1870 changes in policy, 212; corporation ownership, 213; exports, 210; sugar cane in combination with rice cultivation, 213; cane breeding, 12, 142, 213
Jordan, valley of, 24

Labat, J.P., 62, 105, 106, 107, 109

labor: Anguillans, 173; Annamese, 129; Antiguans, 173, 175; Australian Aborigines, 230; Barbadians, 130, 175; British, 230; Canarians, 173; Chinese, 176, 214, 230, 232; Comoro Islanders, 222; European immigrants, 124–5, 223; Filipinos, 224; free Africans, 125–6; Haitians, 167, 173; Indians (Andean), 130, 187; Italians, 230, 232; Jamaican, 167; Japanese, 129, 181, 223, 230, 232; Javanese, 129, 230; Kittitians, 130, 173, 175; Korean, 224; Madeirans, 125, 175–6; Maltese, 230; Mozambiquans, 222; Nevisians, 173; Polynesians ("Kanakas"), 129, 231–2, 229–30, 232; Portuguese, 223, 230; Puerto Ricans, 173; Russians, 224; Singhalese, 230; Somalis, 222; Spanish, 129, 166–7, 224; Tonkinese, 222; Virgin Islanders, 173, Yemenis, 222; *see also* cane-farmers; labor (indentured); *métayage*; slavery; slaves; and *under* individual countries and colonies.
labor (indentured)
    Chinese: abuses of, 127–9; *Nouvelle Penelope* affair, 128; in Cuba and Peru, 128–9; *see also* Cuba; Peru
    Indian, 126–7; *see also* British Guiana; Fiji; Jamaica; Malaya; Mauritius; Natal; Réunion; Trinidad
    Japanese, 129, 181; *see also* Peru
    Pacific islanders, 201; *see also* Australia; Peru
Laet, Jean de, 62
lavrador, 71; *see also* Brazil
Leslie, Eliza, use of sugar in baking, 8
Lesser Antilles, *see under* individual islands
Levant, *see* Mediterranean sugar industry
Ligon, Richard, 77, 81, 88, 98, 109
Lima, Peru, 85, 89
Lippmann, Edmund Oskar, von, 23, 33, 53, 73, 74
livres (French), xiii
Louisiana, *see* United States of America

Madeira: beginnings of industry, 32, 50; early exports to western Europe, Chios, Constantinople, 43, 51–2; cultivation of sugar and land tenure, 52–3; *levadas* (irrigation), 50; number of mills, 53; *alçampremas* (mills), 53; water mills, 53; slavery, 53–4; role of Genoese, 54; decline, 54; relict industry, 54–5
Madeirans: in Brazil, 72; in British Guiana, 125, 175–6; Canaries, 57; Trinidad, 175–6

Magadha, India, Chinese trade mission to, 22
Madagascar: early diffusion of sugar cane to, 24; slaves in Mauritius and Réunion, 221; French annexation 219
Malaya: British annexation and policy, 213–14, 215; Chinese technology 214; Indian labor, 214, 215; crisis of 1880s, 214; Chinese participation, 214–15; British corporations, 215; competitition from rice, rubber, and decline, 215–16
Malta, 34, 230
maltose, 2, 22, 203
manna, 2, 22
manure: around Mediterranean, 36; unused in early American industry, 63; use in eighteenth-century America, 91, 93, 99–100
maple sugar, 2
Marqueses del Valle, 109
Martin, Samuel, 90, 91, 94; recommendations on manure, 100
Martinique, 62, 78, 82, 86, 98, 144; windmills on, 102; competition with beet, 156; post-abolition labor, 156–7; early introduction of central factory system, 157–8; ownership of industry, 158; future, 236
Masters and the Slaves, The, 140
Mauritius: "failure" of Bourbon cane, 141, 219; advantages over Réunion, 219; markets in South Africa, 219; slaves, 221; indentured Indians, 221–2; cane-farmers, 224; central factories, 224–5; *métayage*, 224; research, 228
Mediterranean sugar industry: diffusion of industry to, 23–7; sources for, 32–3; beginnings of industry in, 33–4; northern limits of cultivation, 31–2, 36, 43; cultivation of sugar cane, 36–7; manufacture of sugar, 39–40; refineries, 40, 43; mills and presses, 37–9; fuel, 39; land tenure and labor, 41–3; origins of the plantation system, 42–3; introduction of slavery, 42; technological superiority of western Mediterranean, 38, 73; sugar from Brazil, 45; decline of industry, 43–7
Mesopotamia, early cultivation of sugar cane in, 24
métayage, 156, 224–5
Mexico: regional industry, 85; irrigation, 103; comparison with St. Domingue, 185; expansion from 1870s to the revolution, 183–5; Porfírio Díaz, 184; production, 85, 184; effects of revolution, 185

Mexico City, 85, 95, 100
Middle East, diffusion of sugar cane through, 23–7
Mintz, Sidney, sugar as a drug food, 7, 9
molasses, 107, 108, 109; as sweetener in biscuits and baked beans, 110
Monserrat, 82
Morelos, *see* Mexico
Morocco, 31; establishment of industry, 33–4; 'clayed' sugar, 40; decline, 45–6; *see also* Mediterranean sugar industry
Moslems, *see* Arabs
Muffett, Thomas, 6, 52
must, 1

Natal: market in South Africa, 219; indentured Indians, 222; central factory system, 225; Uba cane replaces Green Natal, 225; research, 228
Nederlandsche Handel Maatschappij, 213
Needham, Joseph, 204, 205
Nevis, 82, 86, 98, 144, 154; density of population, 113; inability to adopt central factory system, 155
New South Wales, *see* Australia
Nile, River, and delta of, *see* Egypt
noria (water wheel), 27

Oman, diffusion of sugar cane through, 24
Otaheiti cane, *see* sugar cane (cultivated varieties)

Palestine, 31, 33; beginnings of industry 34; decline of industry, 43; *see also* Mediterranean sugar industry
passum, 2
Penang, *see* Malaya
Pernambuco, *see* Brazil
Pernambuco and Paraiba Company, 112
Persia, *see* Iran
Peru: three-roller mill, 39, 74; colonial production, 85; irrigation, 103; competition with beet, 180; nineteenth-century production, 180–1; War of the Pacific, 181; capital from guano trade and foreign investment, 181–2; Indian (Andean), Chinese and Japanese labor, 128, 129, 130, 181; comparison with Brazil, 182; Grace and Co., 182
Philippine Islands: minor role of Spain, 216, nineteenth-century exports, 216; Chinese technology, 216; types of sugar, 216; expansion of Luzon, Negros and Cebu, 217; British interest in, 216, 217; annexation by U.S.A., 198, 216, 217; U.S.A. capital in, 217
piculs (Singapore), xiii

264 Index

pikols (Dutch East Indies), xiii
plantations: origins, 42–3; *see also* Americas; and *under* individual countries and colonies
Plat, Sir Hugh, 6
Province Wellesley, *see* Malaya
Puerto Rico, 144, 162; earliest years, 70; population, 113; cheap land in, 163; *colonato* and *colonos*, 170; slavery, 168, 169–70; nineteenth-century development, 168–70; U.S.A. annexation and investment in, 170

qanât, 27
Queensland, *see* Australia

railways, 138; *see also under* individual countries and colonies
raisin wine, 2
ratoon, ratoon cane, ratooning, 13, 56, 91
Réunion: deteriorating position in French Empire, 219; slaves in, 221; indentured Indians in, 221–2; other sources of labor, 222; central factories, 224–5; *métayage*, 224–5; railways in, 228; research, 228; future, 236
Rhodes, 34
Rio de Janeiro, *see* Brazil
Royal African Company, 112
rum: competition with brandy, 110; Royal Navy issue, 110; as *aguardiente, cachça* and *pinga*, 182
run-away slaves, *see* slaves

saccharin, 5
sagwire palm (*Arenga saccharifera*), sugar from, 2, 22
St. Christopher, *see* St. Kitts
St. Domingue: French annexation, 82; beginnings of industry, 86; bagasse for fuel, 99; irrigation, 102–3; population of, 113; maroons, 117; slavery, 115; slave revolt and destruction of industry, 122, 144, 145–6, 185, 189; *see also* Haiti
St. Eustatius, 80
St. Kitts, 62, 78, 82, 86, 98, 130, 144, 158; density of population, 113; abolition 154; railway, 155; technical improvements and central factory, 154–5; future, 236
St. Lucia, 86, 112
St. Vincent, 86, 96, 112
Santo Domingo, *see* Hispaniola
São Paulo, *see* Brazil
São Tomé: as center of innovation, 58, 59, 60; beginnings of industry, 59; mills, 59; quality of sugar, 59, 61; slaves, 59–60; decline, 60–1

sapa, 2
Sasanian empire, 24
senhor de engenho, see Brazil
Sephardic Jews, 79–80, 83, 110
Sicily, 27, 31; beginnings of industry, 34; three-roller mill, 39; *see also* Mediterranean sugar industry
Silk Road, 22
Sind, 25
slavery (abolition of): abolitionist movement, 121–2, 123–4; problem for planters, 123–41; move to unoccupied lands, 124, 214–15
slaves
   food for, 90, 93, 94, 103–4
   numbers of, 114, 115–16, 123–4; on Barbados, 81, 115; in Brazil, 72, 115, 116; on the Canary Islands, 57; on Cuba, 115; in British Guiana, 116; on Hispaniola, 66, 69; on Jamaica, 115, on Madeira, 53–4; Malagasy slaves, 221; on Mauritius, 221; around the Mediterranean, 42; in North America, 115, 116; on Réunion, 221; on St. Domingue, 115; on São Tomé, 59–60
   run-away: *cimmarrones*, 68–9; maroons, 117; Palmares, 117
   value of, in relation to land, 89
   see also labor; labor (indentured); and *under* individual countries and colonies
sorghum (*Sorghum saccharatum*), 2–4; syrup, 4
South Porto Rico Sugar Company, 173
Spain: establishment of industry, 34; center of innovation, 33; decline of industry, 46; surviving fields of sugar cane, 31; *see also* Mediterranean sugar industry
Spanish investment: in Cuba, 168; lack of, in Philippines, 216, 217
Speciale, Pietro, 39, 73, 74
statistics, xiii
Sucreries Coloniales, 224
sucrose, *see* sugar
sugar: properties of, 5; increasing use of in food and drink, 1, 5–9; per capita consumption of, 1, 9, 10, 197, 199, 208, 236; as a drug food, 7; price of, 7, 120, 130, 238–9; influence of low prices in 1883–5, 130, 148, 150, 152, 167, 170, 173, 179, 213
sugar beet, 5, 10, 23, 120, 130, 156, 180, 192, 193, 219, 234, 235, 236; world production of, 234–5
sugar beet industry: origins and development of, 130–3; competition, bounties and tariffs, 133–4, 142, 156, 180, 192, 219; influence on technology of sugar cane

industry, 134–5, 136, 138, 157; effects of
   First World War on, 235
sugar cane
   genus *Saccharum*, 11–13
   cultivation of: methods, 13–14, 16;
      climatic requirements for, 14–15; soils,
      15, diseases and pests of, 15–16, 91, 141,
      142; breeding new varieties of, 16,
      141–2; fire damage to, 91
   introduction to the Americas, 61–2
   possible American origin of, 62–3
sugar cane (cultivated varieties)
   Bourbon or Otaheiti: botanical description
      of, 12; diffusion of, 96, 104, 141, 142,
      147, 150, 164, 189; advantages over
      Creole, 96; "failure" of, 141
   Cheribon, 12, 141, 142, 147
   Creole: botanical description of, 11–12;
      taken to America, 61–3; no alternative
      to, 92; comparison with Otaheiti cane,
      96; replaced by Otaheiti cane, 96, 141,
      142
   Green Natal, 225
   "noble" canes, 12, 141
   POJ (Proefstation Oost-Java) canes, 213
   Ribbon (Batavian stripped, Black Java),
      189
   Uba, 225
sugar cane industry, brief descriptions of,
   16–18, 234–7
sugar (manufacture of)
   general characteristics, 16–17
   Traditional methods: in India, 19–21, 198;
      around the Mediterranean, 39–40; on
      São Tomé, 59; in Hispaniola, 67–8; in
      early Brazil, 76–7; in eighteenth-century
      America, 105–10; in China, 21–2, 206–7
   nineteenth-century innovations: vacuum
      pan, 136–7; centrifugal, 137–8; central
      factories (*usinas*) and central factory
      system, 139–40
sugar mills and presses
   alçaprema, 53, 57, 59
   beam press, 37–8
   edge-runner, 37, 67
   engenho, 71
   ingenio, 65–6, 67
   mortar and pestle, 198, 200; *kolhu*, 202
   screw press, 38
   three-roller mill, 39, 69, 207; origins and
      advantages of, 73–6, 97, 197;
      improvements to, 105; not adopted in
      India, 202
   trappeto, 74
   trapiche, 65, 66
   two-roller mill: origins of, 206; in China,
      75, 205, 207; in India, 198, 199

steam mills, 119, 135
windmills, 38, 39, 102
western technology replaces Chinese, 217
development of modern mill, 135–6
sugar palm, see sagwire palm
sugar (types of)
   non-centrifugal; *azúcar de espumas*, 109;
      batido, 108; "brown sugar," 22, 206;
      clayed sugar, origin of, 108, in
      Americas, 108, in China, 206, in
      Hispaniola, 68, in India, 198, in
      Morocco, 40, in Spain, 108, in
      Philippines, 216; *guḍa*, 20, 21; *gur*, 17,
      19, 20, 198, 206, 234; *jaggery*, 198;
      khanda, 20; *khandsari*, 198; *macho*, 108;
      muscovado, 108, 109; *panela*, 109, 139,
      182, 234; *pilon*, 216, 234; *śarkarā*, 20,
      21; "stone honey," 22; "sugar frost,"
      22, 206; *rapadura*, 139, 182, 234;
      refined, 109; "rock" sugar, 206; world
      production of, 234–5
   centrifugal: definition of 137, 139; replaces
      non-centrifugal, 183; world production
      of, 10, 234
Sung Ying-Hsing, *T'ien-Kung K'ai-Wu*, 204
Syracuse, Gate of the Sugar Workers, 31
Syria, 33; decline of industry, 43
syrup: date, 1; fig, 1; sorghum 4; sugar cane,
   22

Taiwan, 208; from mainland China, 208;
   Dutch in, 209; pre-Japanese annexation,
   209–9; Japanese reorganization, 209
three-roller mill, see sugar mills and presses
Tigris, valley of, 24
Tonkin: early industry in, 21–2; see also
   centers of innovation
trapiche, see sugar mills and presses
Trinidad, 130, 144, 162, 173; exports, 162;
   policies towards ex-slaves, 173–5;
   migrant labor, 175; Madeiran and
   Chinese labor, 175–6; indentured Indian
   labor, 126–7, 176–7; cane-farmers,
   177–8; sources of capital, 178–9; central
   factory system, 179; railways, 179;
   change in ownership of plantations,
   179–80
Tucumán, *see* Argentina

United States of America
   climatic limitations: Florida, 188, 189;
      frost risks, 189; Thomas Spalding
      experiments, 189; South Carolina, 189
   protected industry, 188
   capital in foreign sugar industries, 163,
      167–8, 170, 173, 217

United States of America (*cont.*)
  Hawaii: growing dependence on the U.S.A., 219, 221; annexation, 221; early dependence on Hawaiian labor, 223; immigrant labor, 223–4; agents, 226–7; modernization, 227–8; role of Claus Spreckles, 227–8; research, 228
  Louisiana, 144; benefits of revolt in St. Domingue, 189; Lousiana Purchase, 8, 189; varieties of cane, 189; location of plantations, 189–90; slavery, 190; effects of Civil War, 190–1; post-Civil War labor, 191; central factory system, 191; corporate ownership, 191–2; declining share of U.S. market, 192–3; competition with cotton, 190, and beet, 192; competition with other cane growers, 192; experimental station, 192
usinas, see sugar (manufacture of); and *under* central factories in individual countries and colonies

vacuum pan, *see* sugar (manufacture of)
Venice, role in Crusader states, Crete, Cyprus, 41, 42, 44–5
Verlinden, Charles, 73
Vieira, Father Antônio, 107

water power: early use of and possible role of Crusaders in, 39; on Hispaniola, 66, 67; on Madeira, 53; on São Tomé, 59; in Brazil, 73
Watson, Andrew, 23, 24
West Indies Sugar Company, 173

Yemen: early industry in, 24; Yemeni laborers, 222

Zanzibar, early industry in, 24
Zuazo, Alonso, 63, 65